Finding Our Niche:
The Human Role in Healing the Earth

Finding Our Niche:
The Human Role in Healing the Earth

Donald G. Kaufman

Karla Armbruster

HarperCollins*CollegePublishers*

Finding Our Niche: The Human Role in Healing the Earth

by Donald G. Kaufman and Karla Armbruster

Copyright © 1993 HarperCollins College Publishers

ISBN: 0-06-500771-9

92 93 94 95 96 9 8 7 6 5 4 3 2 1

TABLE OF CONTENTS

PREFACE

The idea for this book came to us over time as we worked on the manuscript for *Biosphere 2000: Protecting Our Global Environment,* an environmental science textbook published separately by HarperCollins (Kaufman and Franz, 1992). Almost every chapter of the text was accompanied by a supplement focusing on a particular topic that exemplified the many physical, biological, and social factors presented in the chapter. For example, our chapter on ecosystem function was illustrated by a discussion of the intricate relationships at work in the Chesapeake Bay ecosystem. Although we gave careful attention to the scientific principles pertinent to each supplement, we also stressed the way in which all environmental problems and solutions are inextricably bound up with social, political, and economic issues.

As we created these supplements, we discovered such a wealth of information on Lake Erie, the restored prairie at Fermi National Particle Accelerator, and many other fascinating examples of environmental issues that we found it hard to limit our discussions to an appropriate length. We believed that teachers and students might want access to some of the information which we just couldn't fit in the textbook. For example, teachers in the Lake Erie area might want to go into more depth on this familiar ecosystem, while those especially interested in how politics and culture affect the human relationship with the environment might be interested in even more details on hunger in North America or population growth in Ethiopia. Secondary teachers who teach topics courses may want to use the book to provide their students with a current, useful look at many of today's most pressing environmental issues. Both secondary and college-level teachers may also find the book appropriate for their own use as they strive to remain current on environmental issues. In general, we thought that teachers and students would appreciate a resource that brings additional information and bibliographic resources on the supplement topics into the classroom.

In addition, we realized that people who might never take or teach an environmental science course are developing an interest in the issues represented by our supplements. They include grassroots activists, outdoorspeople, the politically active, and others. In fact, an ever-growing segment of the population is becoming concerned about the state of our environment, and they want to understand concepts such as how acid rain is formed and why nuclear power threatens both natural systems and human health. As people become more aware of environmental problems, they often become more interested in what role they can play in the solutions to those problems.

As a consequence, they are also searching for information on the social and political aspects of environmental issues. We began to realize that our supplements, in an adapted form, could pro-vide the clear, accurate explanations of both the scientific principles and the cultural factors that such an audience requires in order to form educated opinions on some of today's most pressing environmental issues.

We also became aware that colleges and universities are offering more and more courses in a growing number of disciplines that touch on environmental issues in some way. Instructors in social sciences, business, writing, and many other fields have increasingly realized the importance of highlighting the political and social issues which affect and are affected by not only what goes on in the classroom, but also by what students do after they leave school. As a consequence, instructors throughout many educational institutions have a need for discussions of environmental issues which do not require extensive scientific background and yet do not oversimplify either the scientific or cultural factors involved. Again, we believed that our supplements could easily be adapted to fill such a need.

Because the topics and approach of our supplements seemed in many ways ideal for the needs of the diverse audience we came to recognize, we decided to adapt them specifically for that audience in this book. We've added information and bibliographic sources to make each case study an additional resource for teachers and students using *Biosphere 2000*. In addition, we've placed each supplement into the larger context of the ecological principles and environmental issues it represents—each is presented as an example of a greater concern. This way, all readers can see the larger picture of interconnected organisms and ecosystems and how the individual cases of the supplements relate.

The original supplements had their beginning in the fall of 1983, when Don Kaufman taught an environmental science class for honors students at Miami University, Oxford, Ohio. Those students continued to work after the class ended, conducting research on specific natural systems or resources that involved extensive library and field investigations over the next three years. One student traveled to Boston to meet with a leading authority on whales, another headed north to Marathon County, Wisconsin, to study soil and land use, while yet another ventured to the Everglades of South Florida. The reports they compiled on their research formed the basis for most of our chapter supplements in *Biosphere 2000*, and other students conducted research to update and expand the supplements as time went on. These chapter supplements in turn formed the core of the case studies presented in this book. Although we have changed and significantly expanded the studies, we would like to acknowledge here the creators of the original reports: Karla Armbruster, Lisa Bentley, Doug Davidson, Steve Douglas, Michael Fath, Amy Franz, Pam Gates, Molly Grannan, James Johnson, Frank Andrew Jones, Lindsay Koehler, Tracy Linerode, Debbie Lokai, Pam Marsh, Greg Moody, Tom Pedroni, Cheryl Puterbaugh, April Rowan, Marsha Shook, Cynthia Tufts, and Dana Wilson.

The environment—a vast web of interconnections and interdependencies—is a fitting metaphor for our book projects. Just as *Biosphere 2000* benefitted from the talents of many people, this book also represents the interaction of many minds. Many people were involved in updating and expanding the case studies presented in these pages. To the following people, we owe our sincere thanks for sharing their time and talents to help make this book a reality: Lisa Breidenstein, Cecilia Franz, Kate Grady, Wright Gwyn, Andy Jones, Eric Lazo, Greg McNelly, Nancy Moeckel, Cheryl Puterbaugh, Bobbie Stringer, and Lisa Taylor.

PROLOGUE

In ecology, the term niche is used to describe an organism's functional role in a natural system: what it does, how it does it, and how those activities affect the rest of the system. Determining an organism's niche is no easy task; natural systems and their resident organisms are wonderfully complex. The interrelationships among organisms are sometimes surprising, sometimes baffling, and always enlightening. The complexity of natural systems means that it may takes years or even decades to accurately describe how a single species functions within the confines of its environment.

What does the concept of niche have to do with the average person on the street? A great deal, actually. You see, while humans—like all other organisms—have a niche, we are unique among species in a fundamental way. So far as science can tell, humans are the only species that can consciously change and modify its niche. As we look at the problems we have created, from leaking landfills and dirty air to a reliance on oil and coal, it sometimes seems as if our niche is nature's idea of a practical joke that has gone too far. We use materials and energy with impunity, with little regard for natural systems and other species. But the past several decades have brought an evolution in human thinking. Slowly, people are beginning to realize that we can choose a different path, a different niche. We can heal and preserve the earth, and in the process transform ourselves from conquerors to stewards of nature. This book is about that transformation.

The twenty-one case studies that comprise this book provide specific examples of human interaction with the natural world—both the problems that arise from that interaction and the ways that humans can and do manage resources and ecosystems. Chapters 1–4 illustrate the ecological principles that are essential to understanding how the natural world works and how human activities affect it. By examining Lake Erie, the Chesapeake Bay, the Everglades, and three of the world's great whales (the blue, gray, and humpback), readers can gain an appreciation for the structure, function, and development of natural systems, as well as for the interdependence of organisms and the physical environment. They will also better understand how human activities disrupt and degrade natural systems. Chapter 5, which presents a positive look at prairie restoration on the grounds of the Fermi National

Accelerator Laboratory in Illinois, illustrates how ecological principles can be applied to restore and preserve ecosystems.

Chapters 6–10 examine three critical resources that are linked in an especially significant way—population, food, and energy. As we look at population growth in Ethiopia, hunger in North America, the Alaskan pipeline, strip mining in Appalachia, and a trash-burning power plant in Columbus, Ohio, we see that the interrelationships among population, food, and energy are many and complex. Moreover, environmental issues that concern these three resources often dominate our headlines and profoundly affect our lives. Population and food issues cut to the very heart of our humanness, while energy issues lie at the core of our economic activities and social fabric. Like many environmental professionals, we believe that there is an urgent necessity to balance the human population, food, and energy resources. By including three chapters on energy, we ask readers to first explore the environmental issues related to energy consumption and then to look more closely at society's use and reliance on conventional fuels (oil, coal, and natural gas) as well as alternative sources of energy that might reduce or eliminate our dependence on those fuels.

Protecting the environmental triad of soil, air, and water is examined in Chapters 11–13. Wisconsin's Marathon County, Oregon's Willamette River, and New York's Adirondack State Park provide the backdrop for a close look at the problems and issues pertaining to these basic natural resources. A solid understanding of those problems, and the ways in which these resources have been used in the past, can improve our future management of our soils, waterways, and air resources.

Chapters 14–16 focus on resources and products which are critical to industrial societies: minerals, nuclear resources, and hazardous and toxic substances. These resources provide the raw materials which form the basis for our industries and our economy, but when they are poorly managed, they seriously degrade the environment and pose a significant threat to human health. Consequently, managing them wisely is of the utmost importance. Chapter 17, like the three previous chapters, also focuses on the results of social and economic activity. It explores "unrealized resources"—materials which are typically sent to landfills or incinerators. Treating these materials as resources, rather than as solid *wastes*, and relying on source reduction and resource recovery, is essential if we are to achieve a sustainable and sustaining society.

Clean air, clean water, and fertile soil can help to feed our bodies, but it is wildlands, other living creatures, and human artifacts that feed our minds and spirits. In Chapters 18–21, we examine public lands, wilderness, biological resources, and cultural resources. It is only by understanding and appreciating these resources that we will be able to manage them in ways that are truly environmentally sound.

Back to the concept of niche. Scientists have described the niche of only a very few organisms; much remains to be learned by the ecologists of tomorrow, who today are youngsters exploring a forest, marveling over a fossil found in a creekbed, or watching entranced as a great blue heron stalks its prey in shallow waters. Understanding natural systems and the earth's many life forms is a formidable task indeed. With research, and time, ecologists can provide us with the understanding needed to secure a safe and healthy future. But ultimately, the future is in our hands. We cannot leave it to science alone. There is no doubt in our minds that the greatest challenge to face humankind in the decades ahead is the redefinition of the our human role in nature. We all must take part in that redefinition. We urge you to think of this book as part history, part natural science, and part cultural diary—a record of our past interactions with nature and our recent search to find our niche as healers of the earth.

CHAPTER *1*

ECOSYSTEM STRUCTURE: LAKE ERIE

Discussions of environmental issues often include the term "ecosystem," and many of us have heard the term used to refer to a particular forest, wetland, or lake. But what does the word "ecosystem" really mean? Environmental scientists define an ecosystem as a self-sustaining, self-regulating community of organisms interacting with their physical environment within a defined geographic space. We can gain a valuable perspective on the environment by examining individual ecosystems because they are limited geographically and yet represent the complex web of relationships among living and nonliving components which are the focus of the science of ecology.

To understand an ecosystem, we need to understand more than the living and nonliving components that make up its structure. We might think of this structure as an alphabet, a collection of letters and words that can be combined in many different ways. Knowing letters and words alone does not allow someone to speak a language, and so it is with nature. Simply knowing how ecosystems are structured will not allow us to understand how they work, how they change, and how they respond to human activities—in short, how to speak the language of nature. To speak the language of nature, we must learn how the structural components of an ecosystem function as part of a greater whole. As with language, it is easiest to learn the new alphabet before we try to put the letters together into words and sentences. Thus, in this chapter, we explore a particular ecosystem—that of Lake Erie—by focusing on its structure, although our discussion includes its function as well. We focus on ecosystem function in the next chapter, which discusses the Chesapeake Bay ecosystem.

Lake Erie is part of the 2000 mile-long Great Lakes system, which contains nearly 20 percent of the world's and 95 percent of the United States' fresh surface water. As part of this chain and as a resource in its own right, Lake Erie played an important role in the settlement and the development of North America. Unfortunately, the role human activity played in the structure of the Lake Erie ecosystem was not benign. Decades of pollution, development, and overuse upset the delicate balance among the components of the ecosystem, finally resulting in symptoms such as massive mats of algae and the near eradication of the lake's native fish population. Although a coordinated effort by the Canadian and U.S. governments has helped to repair the worst of the damage, much still needs to be accomplished to restore this complex ecosystem to its natural health.

PHYSICAL BOUNDARIES: THE NONLIVING COMPONENTS OF THE LAKE ERIE ECOSYSTEM

The nonliving, or abiotic, components of an ecosystem include energy, matter (nutrients and chemicals), and physical factors (temperature, humidity, moisture, light, wind, and available space) that affect the density, distribution, and health of organisms. In the Lake Erie ecosystem, the important factors affecting abiotic components include the Lake's position in regard to the other Great Lakes, its size, its depth, its temperature, the amount of sunlight it receives, and the concentration of nutrients in its waters.

The Great Lakes are glacial in origin, formed by the gouging and scraping action of the great ice sheets that once covered much of Canada and the upper Midwest of the United States. The last of these glaciers receded roughly 12,000 years ago. A vital link in the Great Lakes system, Lake Erie accepts water from Lakes Superior, Michigan, and Huron to the west via the Detroit River, and discharges it to the east through the Niagara River into Lake Ontario. The Lake Erie drainage basin has a relatively small ratio of land to water, about 2.5: 1 (most lakes have a land to water ratio of around 6: 1). Hence, the lake is recharged largely through precipitation rather than groundwater from the land surrounding it. Of the groundwater drainage that does contribute to Lake Erie, roughly 70 percent comes from the U.S. side of the lake. Consecutive years of higher-than-average rainfall, such as those that occurred in the early to mid 1980s, contribute significantly to higher water levels in the lake. High waters erode shorelines, flood low-lying areas, and damage homes and businesses built too near the shore or on flood plains.

Although Lake Erie is the fourth largest of the Great Lakes in terms of surface area, it is the shallowest and has the smallest water volume of the chain. It is composed of three distinct basins, making it almost like three lakes in one. The shallow western basin, with an average depth of 24 feet, lies west of Cedar Point, Ohio. The average depth of the central basin, which stretches between Cedar Point and Erie, Pennsylvania, is 61 feet. East of Erie, Pennsylvania, lies the deep eastern basin, with an average depth of 120 feet and a maximum depth of 210 feet.

During the summer, as the sun warms the lake surface, the water tends to form layers. Warm surface waters are less dense than cool, deeper waters. Because the sunny, oxygen-rich upper layer contains a large population of algae that rely on sunlight for their energy, it supports most of the life in the lake. The deepest layer receives less sunlight and supports less life.

Layering rarely occurs in the shallow western basin, where winds and wave action keep the water well mixed. In the central basin, where the bottom layer is narrow, layering poses serious problems. There, the limited oxygen supply is depleted by bacteria that decompose algae and other dead organic material. When the amount of dissolved oxygen in the water reaches zero, the water can no longer support organisms that need oxygen to live. In summer, up to 90 percent of the bottom layer of the central basin can become devoid of oxygen. Because the bottom layer of the deeper eastern basin is thicker and colder, it maintains its oxygen content throughout the summer.

In addition to sunlight and dissolved oxygen, the amount of life that a lake supports depends on nutrients available in the water. Scientists generally agree that the nutrient that acts as the limiting factor in Lake Erie is phosphorus; nitrogen is another important limiting factor.

BIOLOGICAL BOUNDARIES: THE LIVING COMPONENTS OF THE LAKE ERIE ECOSYSTEM

Lake Erie is a place of great beauty and diversity of life. Limestone islands dot the waters of the shallow western basin. Migrating birds and waterfowl, as well as the Monarch butterfly, are regular visitors to the unusual sand spits that reach into the Lake at Presque Isle, Pennsylvania, and Long Point and Point Pelee, Ontario. Quiet wetlands belie a teeming diversity of species. Meanwhile, Lake Erie's waters roar over the falls at Niagara, attracting visitors from around the world.

While Lake Erie's diversity and productivity remain impressive, the living, or biotic, components of the ecosystem have changed dramatically in the past few centuries. When the first Europeans arrived in the Great Lakes basin in the sixteenth century, they found a distinctive biotic community. Smallmouth and largemouth bass, muskellunge, northern pike, and channel catfish favored a near-shore habitat. Blue pike, lake whitefish, walleye, cisco, sauger, freshwater drum, lake trout, lake sturgeon, and white bass flourished in open waters. Mayflies were an important source of food for large fish.

Today, Lake Erie's biotic community is much less diverse. Gone are the blue pike, lake whitefish, lake sturgeon, and cisco. The mayfly, which survives in vastly reduced numbers, is no longer the important food source it once was. Currently, walleye and yellow perch are the prized catch for both sport and commercial fishers. Other numerous fish include white bass, smelt, alewife, sheepshead, and carp; rock bass are fairly abundant around the islands of the western basin.

It is not unnatural for an ecosystem to change; relationships among living and nonliving components are constantly shifting in response to environmental changes. One example of the way a lake ecosystem changes over time is eutrophication—the natural aging of a lake due to high inputs of nutrients, often from natural erosion. The higher level of nutrients stimulates the rapid growth of producer organisms (phytoplankton and algae). In Lake Erie, the primary producer organisms are phytoplankton—organisms that convert sunlight into energy but that also rely on nutrients such as phosphorus and nitrogen. Up to a point, such productivity is desirable, since phytoplankton provide the food base in the Lake Erie ecosystem. Phytoplankton are eaten by zooplankton, which are known as consumer organisms. Other consumers, usually bottom-dwelling fish, eat the zooplankton, and these fish are in turn eaten by fish higher on the food chain. Eventually, the chain can lead to fish-eating birds, mammals like otters and mink, and humans. The gradual accumulation of algae in lakes as they age is one of the signs of natural eutrophication.

However, human activity drastically accelerated eutrophication in Lake Erie, in a process known as cultural eutrophication. Some experts claim the Lake looked twice its natural age in the early 1970s. Massive amounts of nutrients had entered the lake from human sewage and fertilizer washed from agricultural fields. So many producer organisms flourished that vast mats of decaying algae covered the lake. When these organisms died, the bacteria decomposing them depleted the lake's oxygen supply to such an extent that other species in the lake suffered.

While cultural eutrophication influenced the dramatic changes in Lake Erie's species composition, other human activities also played a role. The introduction—intentional and unintentional—of nonnative species drastically upset the delicate relationships among the original biotic community. For example, the sea lamprey, a predator of fish, was accidentally introduced into the Great lakes though the St. Lawrence Seaway and Welland Canal. Although lamprey populations are under control because of the 1950s development of a chemical that killed lamprey larvae without harming other fish, the predator is still considered a problem in the Great Lakes region.

In addition, the ever-increasing flow of toxic substances into the lake influenced species in ways that scientists are still working to understand. Toxics present in the lake or its tributaries may accumulate in the tissues of organisms, eventually reaching a potentially dangerous level of contamination. The higher an organism is on the food chain, the higher the concentration of toxic chemicals in its tissues is likely to be. Because one bottom-dwelling fish such as a rainbow smelt might eat many zooplankton, it will take in all the toxics they have accumulated; a higher level fish such as a trout will in turn eat many smelt and take in all their toxics. As a result of this bioaccumulation, the concentration of chemicals in the tissues of a lake trout can be 1 million times higher than levels in the water. Tumors and genetic deformities are two visible signs of toxic contamination sometimes found in fish and birds.

The adverse effects of human activity are seen in habitats as well as in individual species. Lake Erie was once surrounded by wetlands so extensive that the area around the Western Basin was known as the Great Black Swamp. Today, though, the lake's wetlands are greatly diminished, victims of the plow and urban development. Wetlands are defined as areas where the water table rises above the land's surface for at least part of the year; in addition, the water, when present, is less than seven feet deep and is still or slow moving. Such areas are important because they store water, helping to buffer and control floods. Wetlands also provide habitats for many animal species, including fish that use them for spawning grounds and migratory birds that use them as seasonal habitats and resting places. Finally, wetlands play an essential role in the lake ecosystem, filtering sediments and pollutants from the water, producing oxygen, recycling nutrients, and replenishing groundwater. Given the persistent pollution problems the lake has experienced over the years, it is ironic that so many of its wetlands, the one natural mechanism that can remove pollutants from the lake's waters, have been destroyed.

SOCIAL BOUNDARIES: HUMAN INFLUENCES
ON THE LAKE ERIE ECOSYSTEM

The Great Lakes waterway, which opened the interior of the continent to European penetration and settlement, served as a resource for the settlers that stayed on her banks as well. Drawn by the area's abundant resources, including fish, wild game, and lumber, settlers cleared forested land and drained wetlands for agricultural use. Today, 67 percent of the land in the Erie basin is used for agricultural purposes.

Lumbering became a booming enterprise in the 1870s. Later, iron ore from the deposits near Lake Superior was shipped to Lake Erie ports, and Appalachian coal was sent overland, giving birth to a steel industry. Erie's shores remain home to a heavy concentration of industry. Manufacturing industries, including auto production, steel production, glass manufacturing, and shipbuilding, use nearly 6 billion gallons of Lake Erie water each day.

Approximately 13 million people live in the Erie drainage basin, and 39 percent of the Canadian shoreline and 45 percent of the United States shoreline are devoted to residential use. Six major cities rest on Erie's shores: Windsor, Ontario, Canada; Detroit, Michigan; Toledo and Cleveland, Ohio; Erie, Pennsylvania; and Buffalo, New York. These and other municipalities use lake water for a variety of purposes, including for drinking water and for disposal of sewage and other wastes.

Shoreline development, including marinas, condominiums, and urban revitalization projects such as the Cleveland Lake Front State Park, has a major economic impact on the lake. Tourism and recreational uses of the lake, including fishing, swimming, sunbathing, and boating, are also significant. In recent years, the improvement in the lake's appearance and the resurgence of sportfishing, particularly for walleye, have led to a rising demand for public access to the lake.

Sportfishing in Lake Erie is a multimillion dollar per year industry. The lake is second only to Lake Michigan in sportfishing, and its western basin is known as the "Walleye Capital of the World." The increasing importance of the sportfishing industry affects the lives of a growing number of residents who offer charter boat services, sell sportfishing equipment, and provide accommodations for sportspeople and tourists. Hence, there is increased pressure to manage the fishery in the interest of sportfishers and recreational users.

As the area's economy becomes more service-oriented, commercial fishers often find their livelihood and lifestyle threatened. Competition from the sportfishing industry (particularly in the United States), the decline of valuable species, size limitations, and bans on certain species due to toxic contamination all contribute to the difficulty of commercial fishing. In Canada, the commercial fishery is economically more vital than it is in the United States. The Lake Erie fishery represents over two-thirds of Canada's total Great Lakes commercial harvest. Despite its decline, commercial fishing remains a relatively important industry. The primary targets of commercial fishers are perch, white bass, smelt, and walleye (called pickerel in Canada).

LOOKING BACK: UNDERSTANDING AND CONTROLLING THREATS TO THE LAKE ERIE ECOSYSTEM

For centuries, the people who lived and worked near Lake Erie must have had a sense that the lake's immense size made it virtually immune to damage from pollution and other human activities. In the twentieth century, though, the human population that relies on the Lake Erie ecosystem has realized that the lake is far from immune to such activities. Instead, the lake's ecosystem structure has been seriously damaged by sewage, agricultural and urban run-off, overfishing, toxic chemicals, and other factors associated with human use.

A Growing Awareness of the Plight of Lake Erie

Perhaps the first sign that the pollution and overuse of Lake Erie might have repercussions for its human population came in the early twentieth century with outbreaks of typhoid and cholera. When these outbreaks were traced to untreated human wastes being dumped into the lake, the Canadian and U.S. governments moved to end such dumping. Another result of both governments' new awareness of the lake's vulnerability to pollution was their signing of the Boundary Waters Treaty in 1909. This treaty created the International Joint Commission (IJC) to approve all projects regarding the use, construction on, or diversion of boundary waters, including Lake Erie.

Although the IJC remains an important instrument to preserve and protect the Lake Erie ecosystem today, its power has always been advisory; the IJC does not have the authority to enforce its recommendations. Legislative efforts have been made even more difficult by the need to coordinate various jurisdictional authorities. A Canadian province, four state governments, and the U.S. and Canadian federal governments and numerous local authorities on both sides of the lake must cooperate in order to develop management policy; such cooperation has often been hard to obtain. However, in some areas progress has been made.

Crises of Eutrophication and Overfishing

While the eradication of waterborne diseases and the creation of the IJC signaled a step forward in human management of Lake Erie, other environmental problems went unnoticed until the middle of the century. By this time, municipal sewage was contributing high concentrations of bacteria to the lake, posing a health threat to swimmers. In addition, agricultural run-off, detergents, and insufficiently treated sewage were carrying large amounts of phosphorus into Lake Erie, and the increased nutrient load stimulated cultural eutrophication. By the late 1960s, green plants and algae had flourished to such a degree that seasonal algal blooms spread over entire portions of the central and western basins. Mats of green algae washed ashore, fouling beaches. Newspaper headlines announced "Lake Erie Is Dead," when actually the lake was more alive than ever.

The growing public demand for clean-up action resulted, in 1972, in the Great Lakes Water Quality Agreement (GLWQA), signed by representatives of Canada and the United States. The GLWQA expressed "the determination of each country to restore and enhance the water quality of the largest freshwater system in the world." This precedent-setting agreement established water quality objectives and procedures by which the IJC would monitor the progress of water quality programs. In the years that followed, the two governments spent over $10 billion to reduce levels of pollutants such as phosphorus, oil, and solid wastes. The total amount of phosphates discharged into the lake was reduced 80 percent by upgrading sewage treatment plants and by decreasing the amount of phosphorus allowed in detergents used in the area. As a result, the foul-smelling and unsightly masses of algae cleared up, allowing the concentration of oxygen in the lake to rise and making the Lake Erie waterfront available for tourism again.

About the same time that the high levels of algae were attracting attention, a less visible change was occurring: The lake had lost virtually all its native fish species by 1970. Overfishing had begun to take its toll as early as the turn of the century. By the mid-1950s, the cumulative effect of industrial waste and municipal sewage, agricultural and urban runoff, and drained wetlands and channeled rivers, coupled with the pressures exerted by modern, highly efficient fishing fleets, depleted stocks of many species. In the late summer of 1953, in the lake's western basin, calm, warm weather and a dense algal bloom depleted the oxygen content of the water, and the mayfly population crashed. Because the mayfly population did not fully recover, many popular and valuable fish species followed suit. In 1939, there were 40 commercial fishers working at South Bass Island in the western basin; in 1957, there were none. The 1955 catch of blue pike was 19.7 million pounds. By the decade's end, the blue pike fishery had collapsed; the species is now officially listed as extinct in Lake Erie. The whitefish and cisco fisheries suffered similar fates.

In response, capital expenditures were made for pollution control, especially the redesign of sewage treatment systems. Over a billion dollars was spent by the Cleveland Metropolitan

Sewage Treatment District alone. Agricultural techniques that reduce erosion and the need for chemical fertilization were introduced. Greater protection was given to wetlands, thus preserving vital spawning grounds, and the overharvesting of fish populations was curtailed.

The Insidious Threat of Toxic Chemicals

Although the 1972 GLWQA helped to improve water quality and fish populations in Lake Erie, other, less easily solved problems remained. The same industries that had brought growth and prosperity to the area also for years had discharged dangerous pollutants, such as polychlorinated biphenyls (PCBs), into tributary or Lake waters. Although the discharge of PCBs has been banned, the chemical is still present in the ecosystem. PCBs are organic chemicals that do not degrade easily. They have accumulated in bottom sediments, particularly near harbors and industrial sites. Agricultural and urban runoff continued to contribute pesticides and herbicides as groundwater contaminated by leaking hazardous waste sites continued to seep into the lake.

Symptoms such as fish contaminated with PCBs alerted people that the measures taken since 1972 had not been enough. As a result, the Canadian and U.S. governments adopted the 1978 Great Lakes Water Quality Agreement, a renewal and enlargement of the original GLWQA. The goal of the new agreement was the virtual elimination of discharges of toxic materials into the Great Lakes; the sources it concentrated on were municipal wastewater and industrial discharge. For the first time, management was officially approached on an ecosystem-wide basis rather than by pinpointing symptoms or individual problems.

Another step toward cooperative, basin-wide management took place in 1982, with the establishment of the Council of Great Lakes Governors (CGLG). In 1985, the participants signed the Great Lakes Charter, in which they agreed to consult on, and cooperatively manage, the region's water resources.

Despite the progress made due to prior agreements, 1987 found Lake Erie still plagued with problems that many experts believe stemmed from large amounts of toxic substances. Scientists were disturbed by unusual tumors they found in fish caught in the Great Lakes; tumors were especially prevalent in fish from harbors and river mouths, which are often the areas most heavily polluted. Other scientists reported reproductive failure and birth defects in fish-eating birds such as herring gulls, cormorants, and terns; related problems were also found in mink and otters, species that eat fish. Even the human species seemed affected; one study of mothers on the Ontario side of the lake found that their breast milk contained unusually high levels of PCBs.

In addition to these findings, scientists had recently come to realize that the atmosphere contributed a greater percentage of toxics in the lake than they had previously thought; some reports estimated that over half of the toxic substances deposited in some of the Great Lakes were originally airborne and could be traced to sources such as municipal waste incinerators, automobiles, coal combustion, and pesticides.

As a result of such findings, the U.S. and Canadian governments again came together to review the 1978 GLWQA and passed new amendments emphasizing stricter accountability and management. The new amendments set specific objectives for the reduction of 44 pollutants found in the lake and focused even more strongly than before on pollutants that could not be traced to one specific source, such as agricultural run-off, contaminated sediments, airborne toxics, and polluted groundwater. In addition, the new amendments emphasized coordination of research and the protection of wetlands

Current Problems Due to Introduced Species

Some of the most recent attention Lake Erie has received is in regard to two introduced species: the zooplankton *Bythotrephes* and the zebra mussel (*Dreissena polymorpha*). Both of these exotic species are thought to have entered the Great Lakes system in the bilge water of ocean going vessels, and the zebra mussel in particular is generating increasing concern about its effects on the lake's ecosystem. Named for its bands of black and yellow, the zebra mussel grows up to five centimeters in length and has rapidly spread throughout Lake Erie in the past few years. This temperate freshwater species originated in Russia, but has been common in western and central Europe for centuries. It has flourished in ecosystems around the world so successfully because it can colonize any firm surface and can grow and reproduce very rapidly; one female can produce 30,000 eggs per year.

Although scientists have yet to determine the species' ultimate effect on the lake's ecosystem structure, some changes are already taking place. Because the mussels feed on phytoplankton, they compete with zooplankton, an important food source for fish. In addition, the massive amounts of waste produced by the mussels is decomposed by large numbers of bacteria, which drain the lake's oxygen supply. The mussels pose inconveniences and dangers to the human population as well, clogging intake pipes serving public water supplies, damaging boat engines, and harming fisheries. Although there is no known way to eliminate the zebra mussel population, scientists hope to find ways to control it and its harmful effects on the lake.

LOOKING AHEAD: PROTECTING AND PRESERVING THE LAKE ERIE ECOSYSTEM

Protecting and preserving the unique Lake Erie ecosystem will require a comprehensive management effort that includes measures to ensure water quality, preserve vital habitats, and safeguard the lake's fishery. Specific goals spelled out in the Great Lakes Water Quality Agreement include improving year-round oxygen concentration in the bottom waters of the central basin; substantially reducing the present amount of algae to below that of a nuisance condition in Lake Erie; decreasing erosion rates throughout the Lake Erie basin; controlling the entry of toxic substances into the lakes; developing programs and technologies necessary to eliminate or reduce the discharge of pollutants into the lakes; and restoring and maintaining the health of wetlands so that they are able to perform vital functions.

One threat to water quality in the Great Lakes which remains largely unaddressed is the possibility of oil spills. In 1988, 81 million barrels of petroleum and hazardous materials were shipped through Great Lakes waters; however, in 1989 there were still no plans by oil industry to develop any emergency response centers for spills in the area. Precautions must be taken to protect Lake Erie and the other Great Lakes from the potentially catastrophic effects of such spills.

Maintaining water quality will help to preserve vital habitats and protect the Lake Erie fishery. However, other measures that can help to effectively manage the fishery include standardizing fishing limits and size restrictions between Canadian and U.S. waters. For example, gill net fishing, a method that some people believe leads to overharvesting, is banned in U.S. waters but not in Canadian waters. Fish obviously do not acknowledge the international border; we should be equally unconcerned with it if we are to maintain a thriving fishery and a healthy, balanced ecosystem.

Placing quotas on both sport and commercial fishing will help to eliminate overfishing of desirable species, such as walleye and yellow perch. Currently, quotas are in effect for the West Basin walleye fishery. We can also help to maintain balance in the ecosystem by encouraging increased harvesting of less-desirable species, such as carp, gizzard shad, and freshwater drum, which compete with species already suffering from overfishing.

Because the interrelationships and interdependencies among lake biota are so complex, and our knowledge of them is incomplete, we must protect the health of all organisms in the ecosystem, including forage fish and vegetation. Further, we must protect critical habitats, particularly spawning grounds. Knowledge of the natural system and data derived from long-term research are crucial to the future of Lake Erie and other lake ecosystems.

Ultimately, however, the health of the Lake Erie ecosystem depends upon effective and comprehensive management. Such management would best be accomplished by a consolidated governing body, empowered by the Canadian and U.S. governments to enforce regulations. This is a challenging prospect—the legal, political, and bureaucratic difficulties are many—but the advantages are numerous. Duplication of research, monitoring, and surveillance efforts would be avoided. Interdisciplinary planning, in which specialists from many fields work together to devise management plans, would be far easier to achieve within such a framework.

Although a consolidated management authority is necessary to the improved health of the lake, citizens groups also make a difference throughout the Lake Erie basin. In Ashtabula, Ohio, public participation has played an important role in the development of plans to clean up the Ashtabula River and the city's Lake Erie harbor. Such participation has been duplicated at numerous areas of concern in both the United States and Canada. In Erie, Pennsylvania, citizens have led the effort to clean up the city's bay and harbor-front and to protect the unique habitat at Presque Isle. And at Point Pelee, Ontario, conservationists, concerned residents and sportspeople's groups are taking an active role in the effort to protect and manage Point Pelee Provincial Park.

SUGGESTED READING

Allen, Robert. *The Illustrated National History of Canada: The Great Lakes.* Toronto: McClelland and Stewart, 1970.

Alternatives: Perspectives on Society, Technology and Environment. "Special Issue: Saving the Great Lakes." September/October 1986.

Anderson, D.B. *The Great Lakes as an Environment.* Toronto: University of Toronto Press, 1968.

Ashworth, William. *The Late, Great Lakes.* New York: Knopf, 1986.

Burns, Noel M. *Erie: The Lake That Survived.* Totowa, NJ: Rowman and Allanheld, 1985.

Ellis, W.D. *Land of the Inland Seas: The Historic and Beautiful Great Lakes Country*. Palo Alto: American West, 1974.

Egerton, Frank. *Overfishing or Pollution? Case History of a Controversy on the Great Lakes*. Great Lakes Fishery Commission, Technical Report No. 41. Ann Arbor, MI, 1985.

Evans, Marlene S., ed. *Toxic Contaminants and Ecosystem Health: A Great Lakes Focus*. New York: John Wiley and Sons, 1988.

Farrand, William R. *The Glacial Lakes Around Michigan*. Lansing, MI: Geological Survey Division, Michigan Department of Natural Resources, 1988.

Fortner, Roseanne E., and Victor J. Mayer, eds. *The Great Lake Erie: A Reference Text for Educators and Communicators*. Columbus, OH: Ohio State University Research Foundation, 1987.

Francis, G.R., J.J. Magnunson, H.A. Regier, and D.R. Talhelm, eds. *Rehabilitating Great Lakes Ecosystems*. Great Lakes Fishery Commission, Technical Report No. 37. Ann Arbor, MI: 1979.

Freedman, Paul L., and Bruce A. Monson. "The Great Lakes Water Quality Agreement." *Water Environment and Technology*. October 1989. 285-291.

Gorisek, Sue, and John Fleischner. "A Tolerance for Poison: Plumbing Our Blackest Rivers." *Ohio Magazine*. March 1987. 46+.

The Great Lakes: An Environmental Atlas and Resource Book. Jointly produced by Environment Canada, U.S. Environmental Protection Agency, Brock University, and Northwestern University, 1988. For copies, contact Great Lakes National Program Office, U.S. Environmental Protection Agency, 230 South Dearborn St., Chicago, IL, 60604.

Great Lakes Water Quality Program: Hearing before the Subcommittee on Oceanography, Great Lakes, and the Outer Continental Shelf of the Committee on Merchant Marine and Fisheries, House of Representatives, One Hundred Second Congress, first session, April 30, 1991. Washington, DC: U.S. Government Printing Office, 1991.

Hart, M. "Invasion of the Zebra Mussels." *The Atlantic*. July 1990. 81-87.

Hatcher, Harlan. *Lake Erie*. New York: Bobbs-Merril, 1945.

Havighurst, Walter. *The Great Lakes Reader*. New York: Macmillan, 1966.

Hileman, Bette. "The Great Lakes Cleanup Effort." *Chemical and Engineering News*. February 8 1988. 22-39.

Kuchenberg, Tom. *Reflections in a Tarnished Mirror: The Use and Abuse of the Great Lakes*. Sturgeon Bay, WI: Golden Glow, 1978.

Lake Erie Estuarine System: Issues, Resources, Status, and Management. Proceedings of a seminar held May 4, 1988. Washington, DC: U.S. Dept. of Commerce, National Oceanic and Atmospheric Administration, NOAA Estuarine Programs Office, 1989.

Lewis, Jack. "The Five Sister Lakes: A Profile." *EPA Journal*. March 1985. 5-6.

Makarewicz, Joseph C, and Paul Bertram. "Evidence for the Restoration of the Lake Erie Ecosystem." *BioScience*. April 1991. 216-223.

McCarthy, Anne. *The Great Lakes*. New York: Crescent Books, 1985.

Nowak, Henry J. "The Benefits of a Cleaner Lake Erie." *EPA Journal*. March 1985. 7-8.

Roberts, Leslie. "Zebra Mussel Invasion Threatens U.S. Waters." *Science*. September 21 1990. 1370-1372.

Rousmaniere, John, ed. *The Enduring Great Lakes*. New York: Norton, 1979.

"Saving the Great Lakes." *Environment*. April 1985. 22.

U.S./Canada Great Lakes Water Quality Agreement: Hearing before the Subcommittee on Investigations and Oversight of the Committee on Public Works and Transportation, U.S. House of Representatives, Ninety-ninth Congress, second session, July 30, 1986. Washington, DC: U.S. GPO, 1986.

CHAPTER 2

ECOSYSTEM FUNCTION: THE CHESAPEAKE BAY

In the previous chapter, we explored the Lake Erie ecosystem by focusing on its structure. As that chapter illustrated, an ecosystem's structure is made up of a tremendous variety of living and nonliving components. However, it is process—the movement of energy and materials through the ecosystem—that links these components together as a functional unit. All ecosystems, from the driest desert to the wettest tropical rain forest, are dependent upon the flow of energy and the cycling of materials through the community of living organisms.

Most energy in an ecosystem comes originally from the light of the sun. Green plants and algae capture and convert the sun's light energy to chemical energy, which they store primarily as carbohydrates (sugars or starches) or lipids (fats). Surprisingly, very little of the sun's energy which enters our atmosphere is actually used by living organisms. Forty-two percent of that energy heats the land and warms the air, 23 percent helps regulate the water cycle through evaporation, and about 1 percent generates wind currents. Although a mere 0.023 percent of the sunlight reaching the earth is actually used for photosynthesis, that small fraction results in the hundreds of billions of tons of living matter, or biomass, which covers our planet.

Organisms such as green plants and algae which store energy through photosynthesis are known as either photoautotrophs or producers. Photosynthesis by photoautotrophs is roughly 1-3 percent efficient at converting light energy to chemical energy: one hundred units of light energy would produce one to three units of chemical energy. Photoautotrophs comprise the first, or lowest, trophic level of energy conversion and use in any biological community; organisms at each successively higher trophic level depend upon the organisms at lower levels for energy that they obtain through the food chain. The second level—primary consumers—consists of herbivores, or animals that eat producer organisms. Carnivores, which eat other animals, and omnivores, which consume both animals and producer organisms, make up the succeeding levels. For example, weedy plants are eaten by grasshoppers, which in turn are eaten by meadowlarks, which in turn are eaten by Cooper's hawks.

The linear concept of the food chain tends to oversimplify the flow of energy through an ecosystem, however. It's important to realize that organisms that decompose dead organic matter operate at every trophic level, breaking the matter down into its constituent compounds. Detritus food chains, which are characterized by consumers of dead or decaying materials, are important in both aquatic and terrestrial habitats. In addition, food chains intersect with each other in more complex associations called food webs. Food webs start with producers of many kinds, consumed by many species of consumers at several trophic levels, and culminate with decomposers working at all levels.

Materials from the air, water, and soil also cycle through the ecosystem's food webs and then back to the air, water, and soil. Throughout the earth's history, materials have continued to cycle in this way. For example, nutrients or gases are released into the soil, water, or air when microorganisms decompose once-living tissue to simpler molecules—molecules that are subsequently affected by chemical and physical changes until they are in a form that can once again be used by living organisms. Materials that are of special concern in aquatic ecosystems like Lake Erie and the Chesapeake Bay are phosphorus and sulfur. These nutrients cycle primarily between the land and oceans or lakes and back to the land again (although they may include a gaseous phase). If aquatic ecosystems are overloaded with these nutrients, though, massive algal blooms may occur (see the discussion of cultural eutrophication in Chapter 1), blocking light from submerged vegetation and lowering the oxygen content of the water.

Although every ecosystem is characterized by many complex relationships and functions that cycle energy and materials, many times we do not recognize the importance and interdependence of these relationships until they are threatened. One type of ecosystem that is especially vulnerable to to pollution by toxic materials and to damage from excessive amounts of nutrients such as phosphorus and nitrogen is the estuary, a semi-enclosed coastal body of water composed of fresh and saline (salt) water. Estuary waters support many marine organisms during their crucial spawning and nursery stages, and thus these organisms are exposed to any problems such as toxic substances, lack of oxygen, or lack of food when they are at their most vulnerable. Because of their position between fresh and coastal waters, estuaries tend to receive heavy loads of pollution both from marine waste disposal and from contaminated freshwater. In addition, because estuarine ecosystems are shallow and semi-enclosed, they tend to retain pollutants.

Such problems have surfaced in the Chesapeake Bay, the largest and most productive estuary in the United States. The bay has attracted widespread attention due to dramatic losses in biological productivity which have been traced to pollution, excessive nutrient loading from agricultural run-off, overharvesting of fish and shellfish, and encroachment by human population and development. When the Chesapeake's declining productivity and deteriorating water quality first became widely recognized in the 1970s, many people became seriously concerned about restoring the estuary's ecological health. The history of the Chesapeake Bay ecosystem demonstrates not only the complexity of ecosystem function, but also its vulnerability to overuse and abuse by the human population.

PHYSICAL BOUNDARIES: THE MEETING OF FRESH AND SALT WATER

The 10,000 year old Chesapeake Bay begins near the Susquehanna flats in Maryland. From there, the "Queen of Estuaries" virtually slices the state in half, stretching 190 miles south to meet the Atlantic Ocean at tidewater Virginia. Along those 190 miles, the bay's width ranges between 4 and 30 miles.

Fresh water drains into the Chesapeake from a 64,000 square mile area that stretches over six states—Pennsylvania, Maryland, Virginia, Delaware, West Virginia, and New York. More than

150 rivers, streams, and creeks contribute to the bay. Of the bay's 50 major tributaries, the Susquehanna River, which originates in Pennsylvania, provides half of the bay's inflowing fresh water, and seven others—the Patuxent, Potomac, Rappahannock, York, James, and Choptank Rivers, and the West Chesapeake Drainage Area (the Gunpowder, Patapsco, and Back Rivers)—contribute another 40 percent.

The land use and management practices of the areas through which all these rivers flow determine the volume and chemical characteristics of the fresh water discharged to the bay. For example, the chemical composition and water quality problems of the heavily industrialized Elizabeth and Patapsco Rivers are very different from those of the Choptank or Rappahannock Rivers.

Daily ocean tides, which vary in height and in the degree to which they penetrate the estuary, also affect the bay's physical characteristics such as circulation and salinity. Salinity, a measure of the concentration of dissolved salts in the water, varies throughout the Chesapeake. Salinity is highest at the mouth of the bay, where marine waters enter the estuary, and gradually decreases toward the northern end of the main stem. Salinity concentrations also vary vertically and horizontally: deeper waters and waters on the eastern side of the bay are more saline.

The important physical processes in the Chesapeake are ocean and river current action, and the resulting interaction of salt and fresh water. Because salt water is more dense, it flows beneath the fresh water. Mixing patterns also depend on the physical dimensions of the estuary. Comparatively shallow estuaries such as the Chesapeake Bay have a two-layer flow: An upper layer of fresh water flows seaward, while an incoming layer of salt water flows beneath. Where two layers intersect, fresh water is mixed downward and salt water upward in what is called a vertical mixing zone.

The varying chemical composition of estuarine waters allows for diverse habitats. If an estuary is ecologically healthy, it can be one of the most productive ecosystems in terms of its energy flow and the numbers and variety of its biota.

BIOLOGICAL BOUNDARIES: THE FLOW OF ENERGY AND MATERIALS THROUGH THE BAY'S COMMUNITIES

The Chesapeake: It is an ecologist's paradise, an artist's Utopia, a recreationist's playland. The bay is a place of wonder and great beauty, a natural resource whose bounty has drawn millions to its shores. Think of the Chesapeake and a thousand images come to mind— watermen setting out from their docks before dawn; the rising sun over still waters; a flock of geese high overhead; a flash of movement in tall reeds; children at play on a beach; a heron stalking prey in marsh shallows; a blue crab scuttling among bay grasses. Each of these, and hundreds more, provide but a glimpse of bay life. Alone, they are simply mental snapshots, lovely but incomplete. Taken together, they form a panorama of an ecosystem, the never-ending cycle of life and death that binds all organisms.

Five important, interacting biological communities comprise the Chesapeake Bay ecosystem: marshes or wetlands; plankton; submerged aquatic vegetation, sometimes known as baygrass; benthic or bottom-dwelling organisms; and nekton or swimmers.

Wetlands are vegetated areas that act as a natural boundary between land and water. Kept moist by run-off, groundwater, adjacent streams, and bay tides, wetlands are dominated by marsh plants that provide valuable habitat for other members of the bay ecosystem. Waterfowl, furbearers, and other animals, and the young of commercially-important fish and shellfish depend on wetlands for food and shelter. Marsh plants also filter nutrients from inflowing water, thus decreasing potentially harmful overloads of nutrients to the Chesapeake. Draining and dredging of wetlands to provide land for agricultural and residential use pose a major threat to this vital biotic community.

Plankton, tiny floating organisms that drift with the water's movement, include phytoplankton (microscopic plants), zooplankton (microscopic animals), bacteria, and jellyfish larva. Phytoplankton are producers that occupy the first trophic level in the bay ecosystem. Increased nutrient loading to the bay dramatically increases the amount of phytoplankton, causing algal blooms that threaten other communities in the ecosystem.

Although the bay is home to approximately fifteen different species of submerged aquatic vegetation (SAV), eelgrass is the most prominent. Found only in waters of the estuary shallow enough for light to reach the bottom, SAV species must remain moist and live with their leaves at or below the water's surface. They are a primary food source for a variety of organisms, including herbivores like ducks and Canada geese. They are particularly important links in detritus food chains. For example, many species of invertebrates feed on decaying aquatic grasses. The invertebrates are an important food source for small blue crabs and fish, such as striped bass and perch; these, in turn, provide food for wading birds, including herons. The decomposition of these higher order consumers, as well as other organisms at all trophic levels, returns nutrients to the physical environment, where they are once again taken up and used by SAV.

The SAV community provides habitat for many organisms, including molting blue crabs and spawning fish. In addition, submerged vegetation tends to slow down water velocities and cause particulate matter to settle at the base of their stems, resulting in clearer water; it also acts as a buffer to shoreline erosion. As SAV has declined in the bay, the average rate of erosion along its shore has risen to 10 feet per year in some spots. Another function SAV performs is to take up nitrogen and phosphorus, thus buffering the bay against excess quantities of these nutrients. Ironically, although SAV protects the bay against algal blooms in this way, the algal blooms that do occur pose a major threat to SAV communities by blocking the sunlight they need for photosynthesis.

Benthic communities consist of organisms that live on or in the bottom of the bay outside of the marsh and grass beds. They are composed primarily of such commercially important organisms as oysters, blue crabs, and clams. Like SAV, benthic organisms are adversely affected by nutrient loading and increasing amounts of phytoplankton, as decomposition of dead phytoplankton depletes the oxygen in the bottom waters.

The nekton community is composed of the swimmers of the bay—the fish, certain crustaceans, squid, and other invertebrates. The approximately 200 species of fish found in the bay are classified as either resident or migratory species. Residents include killifishes, anchovies, and silversides. Migratory species, which are generally larger than resident fishes, are those that spawn in the bay or its tributaries and those that spawn in the ocean. The bay is especially well known as the spawning ground for 90 percent of the Atlantic population of striped bass, also known as rockfish.

Catches of freshwater-spawning fish have declined in recent years. From 1880 to 1980, marine spawners accounted for 75 percent of the bay catch, but during the period from 1971 to 1980, the figure was 96 percent. The decline of the freshwater fishery is generally attributed to the pollution or destruction of spawning areas and to the increasing salinity of the Chesapeake Bay.

In addition to these major communities within the ecosystem, twenty-nine species of waterfowl depend upon the bay ecosystem for their habitat. Twenty-four of these species are ducks, two are swans, and three are geese. Over 75 percent of the Atlantic Flyway waterfowl population winter in tidewater areas of the bay.

SOCIAL BOUNDARIES: THE RELATIONSHIP OF THE HUMAN POPULATION WITH THE BAY ECOSYSTEM

The Chesapeake Bay fulfills many important functions for the surrounding human population. Traditionally, area residents have taken great pride in the bay's aesthetic qualities: its clear blue waters, vast wetlands, and abundant wildlife. The bay is also one of the most popular recreational resources on the East Coast. There are 1750 miles of navigable shoreline, and the value of the sportfishing industry is nearly $300 million annually. Swimming and boating are supported by numerous beaches and safe harbors.

A healthy and productive Chesapeake Bay is vital to the region's economy as well. Fishing has yielded an average of 27 million pounds of oysters annually for the past 50 years; the bay is also the largest producer of blue crabs in the world, with yearly harvests of approximately 55 million pounds. The value of the finfish and shellfish harvests is approximately $1 billion annually. In the United States, only two other areas out-produce the Chesapeake Bay—the Atlantic and Pacific oceans.

Land uses in the Chesapeake basin are diverse. The rich soils of the coastal plain support agriculture, large areas of the Piedmont are occupied by forestry uses, and the Eastern Shore hosts poultry, seafood, and vegetable processing industries. Located along the major tributaries are industrial facilities for steelmaking, chemical production, leather tanning, plastics and resin manufacturing, paper manufacturing, and shipbuilding. The economic importance of Chesapeake Bay ports is staggering. During 1979, more than 90 million tons of cargo, worth nearly $24 billion, were shipped via the bay.

Given the beauty and abundant resources of the Chesapeake Bay drainage basin, it's little wonder that the area has experienced a tremendous growth in population. For example, between 1959 and 1980, the population grew by 4.2 million. The region's rising population has contributed significantly to the stresses on the bay ecosystem. Since the mid-twentieth century, the rate at which land has been converted to residential areas has increased by 182 percent. Developed land comes at the expense of wetlands and forests and often exacerbates soil erosion, which can result in the buildup of silt in the bay. Population growth and urban development have brought increased municipal wastewater discharges and have concentrated industry, and thus industrial pollution, in certain areas. Recent studies show that increased air pollution from industry and automobiles can also affect the function of the bay ecosystem.

LOOKING BACK: THE HISTORY OF THE CHESAPEAKE BAY AREA

Even before the first European settlers arrived at the shores of the Chesapeake Bay, it was famed for its biological productivity. Native American inhabitants had christened the bay "Chessepioc," meaning "the great shellfish bay"—a tribute to the huge reefs of oysters that not only provided the natives with food but also gave the bay its clarity by constantly filtering water. The bay also provided seemingly boundless harvests of fish such as shad and herring, and attracted huge numbers of waterfowl.

By the 1970s, however, it had become painfully clear that the bay's resources were not boundless. The oyster population had dramatically decreased; although records from the late 1800s showed harvests of 17 million bushels in one year, the annual average from 1966 to 1980 was between 2 and 3 million bushels. By the 1983-1984 season, the harvest had dwindled to a mere 868,0000 bushels. Fish populations also declined throughout the 1970s and 1980s; for example, commercial catches of striped bass dropped from 14.7 million pounds in 1973 to 1.7 million pounds in 1983. Dams along the bay's tributaries altered fish migration and diminished the amount of fresh water entering the estuary. Today, the bay's SAV covers only about 50,000 acres, a drastic reduction from its original 100,000 to 300,000 acres. This decline has been linked to an increase in phytoplankton, which block sunlight needed by the grasses. In addition, over half of the bay's wetlands and 40 percent of its forests have been replaced by agriculture or development.

As both the public and government agencies became aware of the plight of the bay, they began to ask what forces lay behind its decline and how its health might be restored.

Cultural Forces Responsible for the Decline of the Bay Ecosystem

Although the bay's health and productivity seem to have declined rapidly over the past few decades, this decline actually has been the culmination of centuries of human pressures on the ecosystem's vital functions and components.

Early settlers merely dumped their sewage and waste into the bay or its tributaries, and their agricultural practices led to erosion, which deposited silt in the bay. As the population of the bay area grew, forests were cleared to provide more farmland and agriculture was conducted on an ever greater scale. Siltation became a major problem for fish, navigation, and public health during the 1600s and 1700s. Numbers of herring and shad declined in the 1820s, a development attributed to siltation as well as the action of ship wakes and the damming of streams.

By the early twentieth century, urbanization and an ever growing human population intensified the negative impact on the bay. Sewage and industrial waste from cities in the bay's watershed replaced siltation as the principle water quality problem. Some cities, like Baltimore, Maryland, and Norfolk, Virginia, took judicial and legislative action to address these problems, but on the whole, cities did not assume responsibility for clean-up efforts.

Water quality continued to deteriorate through the middle of this century. Problems included siltation, turbidity, algal blooms, concentrations of toxics and organic compounds in the sediment and water, a decrease in the amount of oxygen in the water, and a general decline of submerged aquatic vegetation. Most of these problems have been traced to two kinds of pollution: point source and nonpoint source.

Until recently, point source pollution—pollution that can be traced to an identifiable source—received most of the blame for the bay's degraded condition. Over 2750 municipal sewage treatment plants and industrial facilities discharge directly into the bay. Some of the toxic substances found on the bottom of the bay in high concentrations are polychlorinated biphenyls (PCBs), kepone, and DDT. Metals such as cadmium, chromium, lead, and zinc are found in the bay's tributary river systems. Once such toxics and metals enter the bay ecosystem, they can harm a wide range of species as they cycle through.

Municipal treatment plants are also the point source of a different and perhaps more far-reaching type of pollution. Discharge from sewage plants contains phosphorus and nitrogen. Increased levels of these nutrients in the bay encourage the growth of phytoplankton, especially blue-green algae, with a resulting lack of light and oxygen which harms both SAV and benthic organisms. The areas of the bay suffering the heaviest loads of nutrients are the Patuxent, Potomac, and James Rivers in the east, and the north and central main bay.

Although point source pollution has been blamed for many of the bay's problems in the past, it is becoming increasingly apparent that nonpoint pollution is also detrimental to the ecosystem. Nonpoint sources of pollution, including agricultural and urban runoff, have a significant impact on the Chesapeake because the bay's drainage basin is so large. Farms and cities hundreds of miles away in a number of states contribute pollution to the bay. Nonpoint pollution is difficult to control not only because its sources are hard to pinpoint exactly, but also because their widespread nature makes coordination of any management efforts complicated. All told, nonpoint sources contribute 39 percent percent of the bay's excess phosphorus and 67 percent of its excess nitrogen.

Throughout the bay's drainage basin, ongoing loss of wetlands contributes to nonpoint pollution. In Maryland, agricultural drainage is the primary reason for wetland loss; in Virginia, channelization projects, especially for agricultural purposes, are the principle cause. Residential development, industrial projects, expansion and development of marinas, and dredge-and-fill activities also play a role. The additional farmland provided by drained wetlands adds to the flow of nutrients and pesticides to the bay. In addition, the loss of wetland areas means fewer of these nutrients and pesticides are removed by wetland vegetation before they reach the bay.

Efforts to Restore the Health of the Bay Ecosystem

In response to the disturbing state of the Chesapeake, the Chesapeake Bay Program (CBP) was instituted in 1976. Coordinated by the EPA, the CBP is our country's oldest estuarine management program. The CBP was designed to coordinate the efforts of many diversified groups—local , state and federal agencies, as well as public and independent groups such as colleges and universities—in a single cooperative effort. Its four main objectives were to describe historical trends in the bay ecosystem; determine the current state of the bay by evaluating ongoing research; project future conditions of the bay; and identify alternative strategies for managing bay resources.

CBP participants identified ten critical management areas. The decline in SAV, eutrophication, and toxic accumulation in the food chain were given high priority. Medium priority issues included dredging and disposal of contaminated sediments and control of shellfish and fish harvests. Wetland alteration, shoreline erosion, and the water quality effects of boating and shipping were considered important, but lower, priorities.

The CBP immediately initiated a comprehensive, seven-year study of the bay which took the unprecedented approach of characterizing the bay as an integrated ecosystem. It acknowledged the differing needs and problems of various sections of the bay and recommended that control strategies be targeted by geographic area. Perhaps most importantly, the CBP recognized the need for a bay-wide management authority to coordinate the activities of the various federal and state planning and regulatory agencies.

Another result of the CBP study was the 1983 Chesapeake Bay Restoration and Protection Plan, which identified the bay's most important problems, assessed current pollution control efforts, and set general goals for abatement of both point and nonpoint source pollution. The Plan established an overall goal of restoring the bay to the conditions it enjoyed in the 1950s.

In addition, that same year the governors of Virginia, Pennsylvania, and Maryland, the mayors of the District of Columbia, the EPA administrator, and the Chesapeake Bay Commission Director issued the Chesapeake Bay Agreement of 1983. This Agreement established the Chesapeake Executive Council to oversee the variety of programs designed to improve and protect water quality and living resources of the bay. Participants in the 1983 Agreement established bay-wide goals and objectives for controlling nutrients and toxics, protecting and restoring the bay's living resources, habitats, and ecological relationships, managing all related environmental programs with a concern for their impact on the bay, and working for cooperation among all levels of government.

As of 1987, progress toward these goals included a $100 million effort to control nutrient loading in order to diminish algal blooms and raise the concentration of oxygen in the water. Other actions designed to reduce nutrient loading included a ban on phosphates in Maryland and the District of Columbia. In addition, Maryland, Pennsylvania, and Virginia have all instituted new programs to control nonpoint source pollution, and municipal sewage treatment plants throughout the bay drainage basin have instituted measures to reduce levels of pollution in their discharge.

Progress in other areas of concern includes attempts to control and limit development near the bay and its tributaries as well as the ongoing restoration of areas once covered with SAV. The declining fish population is being addressed through a moratorium on harvesting striped bass as well as the installations of fish lifts, ladders, and dams to help fish reach the bay from rivers in Virginia and Maryland. Most recently, in 1991, the CBP published a guide to assist federal and state agencies in protecting and restoring wetlands.

Although such government and institutional programs are essential to the protection and improvement of the Chesapeake Bay ecosystem, efforts of individuals and citizens' groups have also proven invaluable. Since 1985, trained citizen monitors for the Alliance for Chesapeake Bay (a federation of citizens' organizations, businesses, scientists, and individuals) have been studying numerous near-shore sites along tidal tributaries of the bay, such as the James and Patuxent Rivers. They collect data on water quality factors like the amount of dissolved oxygen in the water, salinity, pH, clarity, and temperature. Their work, which supplements a baywide monitoring program begun in 1984, has yielded some important insights. For example, the shallower near-shore habitats have greater swings in the levels of dissolved oxygen than do the deeper waters of the rivers and streams (which are closely watched as part of the baywide monitoring program). Species in these habitats are thus subjected to greater stress than those in deeper waters. The Chesapeake Bay Foundation, a nonprofit organization, has initiated educational and land acquisition projects as well as taken part in legislative, administrative, and judicial proceedings affecting the bay. The dialogue between these citizens' groups and researchers and policymakers has been invaluable in protecting the bay and must be maintained.

LOOKING AHEAD: THE FUTURE OF THE CHESAPEAKE BAY ECOSYSTEM

The Chesapeake Bay Program has been considered markedly successful in its attempts to discover and control the sources of the bay's problems. This success has been attributed to a variety of factors: adequate preliminary research into the conditions and trends of the bay and the sources of pollutants; adequate funding (nearly $30 million was spent on the initial 7 year study); a long-term approach which spent enough time collecting data, defining problems, and establishing institutional relationships); and strong public participation.

Although the groundwork laid by the CBP and the organizational framework established by the 1983 Chesapeake Bay Agreement have made great strides towards the wiser management and restored health of the bay, many steps remain to be taken. Efforts to reduce nutrient loading from agriculture and sewage must continue to involve not only those states bordering the bay, but all areas in its watershed. In addition, the Chesapeake Bay Foundation has called for improved measures designed to protect water quality: new or revised water quality standards for known toxic chemicals, nitrogen controls for sewage plants located in the most highly enriched tributaries, increased funds to encourage farmers to practice conservation measures, and bans on phosphate detergents in Pennsylvania and Virginia.

In addition to implementing measures to protect and improve water quality in the bay, the CBP should continue research into the many important questions remaining in regard to the ecosystem's health. Researchers are still unsure what specific nutrient reduction levels need to be met to assure water quality and to protect living resources of the bay. Many species that are especially vulnerable to the bay's problems still need to be identified and protected. The specific conditions necessary to protect and restore living resources in each area of the bay still need to be determined. Although regulators have concentrated on the role of phosphorus in nutrient loading because it is more crucial than nitrogen to algal growth in fresh water, the role of nitrogen in the saline water of the estuary remains uncertain. Continued research should provide the answers to these and other questions. As more information becomes available, and our knowledge of this intricate ecosystem grows, management of the Chesapeake Bay should improve.

Finally, those concerned with the bay's future need to address the root of the problem: too many people altering too many parts of the ecosystem. Between 12 and 13 million people currently live in the Chesapeake Bay drainage basin. That figure is expected to grow to 14 million by the year 2000, and to 15.6 million by 2020. Not surprisingly, the amount of developed land in the area is predicted to expand 60 percent by 2020. As we have seen, it is the impact of human activities which is endangering the bay ecosystem, and continued growth of the human population of the bay area is certain to place increasing stress on the ecosystem. Development must be managed so as not to disturb the ecosystem's function if we are to restore and maintain the ecological integrity and productivity of this invaluable resource.

SUGGESTED READING

Arnold, Joseph L. *The Baltimore Engineers and the Chesapeake Bay, 1961-1987*. Baltimore, MD: Baltimore District, U.S. Army Corps of Engineers, 1988.

Baird, Daniel, and Ulanowicz, Robert E. "The Seasonal Dynamics of the Chesapeake Bay Ecosystem." *Ecological Monographs*. December 1989. 329-364.

Carr, Lois Green, Philip D. Morgan, and Jean B. Russo, eds. *Colonial Chesapeake Society*. Chapel Hill, NC: Published for the Institute of Early American History and Culture by the University of North Carolina Press, 1988.

Chesapeake Bay: Introduction to an Ecosystem. Washington, DC: U.S. Environmental Protection Agency, 1982.

Cooper, Sherri R., and Grace S. Brush. "Long-term History of Chesapeake Bay Anoxia." *Science*. November 15 1991. 992-996.

Dybas, Cheryl Lyn, and Lynda Richardson. "Homeward Rebound: The Osprey's Dramatic Return to the Chesapeake Bay is Due in No Small Part to the People Who Live There." *National Wildlife*. April/May 1990. 42-48.

Horton, Tom. *Bay Country*. Baltimore: The Johns Hopkins University Press, 1987.

——, and William M. Eichbaum. *Turning the Tide: Saving the Chesapeake Bay*. Washington, DC: Island Press, 1991.

Martin, S.O. "Last Chance for Chesapeake Bay." *Planning Magazine*. June 1986. 12-19.

McCloskey, William. "Along the Chesapeake: Heaven and Earth Agree." *Oceans*. May/June 1985. 3+.

Morales, Leslie Anderson. *The Chesapeake Bay Cleanup, Model of Inter-Jurisdictional Cooperation: A Bibliography*. Monticello, IL: Vance Bibliographies, 1989.

Officer, C.B., et al. "Chesapeake Anoxia: Origin, Development, and Significance." *Science*. January 6 1984. 22-27

Orth, R.J., and K.A. Moore. "Chesapeake Bay: An Unprecedented Decline in Submerged Aquatic Vegetation." *Science* 222: 51-53, 1983.

Powers, Ann. "Protecting the Chesapeake Bay: Maryland's Critical Area Program." *Environment*. May 1986. 5+.

Sun, Marjorie. "The Chesapeake Bay's Difficult Comeback." *Science*. August 15 1986. 715-717.

Tate, Thad W., and David L. Ammerman, eds. *The Chesapeake in the Seventeenth Century: Essays on Anglo-American Society*. Chapel Hill, NC: Published for the Institute of Early American History and Culture by the University of North Carolina Press, 1979.

U.S. Congress, Office of Technology Assessment. *Wastes in Marine Environments*, OTA-O-334. Washington, D.C.: U.S. Government Printing Office, April 1987.

Warner, William W. *Beautiful Swimmers: Watermen, Crabs, and the Chesapeake Bay*. Boston: Little, Brown, 1976.

CHAPTER 3

ECOSYSTEM DEVELOPMENT AND DYNAMIC EQUILIBRIUM: THE EVERGLADES

We often like to think that if nature is left undisturbed, it reaches and maintains a balance. However, it would be inaccurate to think of this balance as a static state. In fact, an ecosystem can persist through time only by constantly changing in reaction to environmental fluctuations and thus maintaining what scientists call a dynamic equilibrium or a dynamic steady state. Each organism, group of organisms, ecosystem, and even the earth itself are all subject to constant environmental fluctuations such as variations in rainfall, temperature, and sunlight. By adapting to these variations, an ecosystem changes and yet preserves its long-term stability; thus, a stable ecosystem is one that maintains a dynamic equilibrium over time.

Ecosystems, and ultimately the earth itself, maintain a dynamic steady state in one of two ways: either by resisting changes so extreme that they could endanger long-term stability or by restoring structure and function after a disturbance. An ecosystem's capacity to resist or restore depends upon the intricate interrelationships among organisms and their physical environment. For example, in a stable ecosystem, populations of different organisms balance each other's inputs and outputs. Energy inputs from the sun become a plant's outputs when it is consumed by animals. Animals' outputs, in the form of organic substances or waste, become the inputs for other animals. Another type of interaction that helps to maintain the health and stability of an ecosystem is the competitive relationship existing between predator and prey. Although predators need the prey for a food source, the prey also rely on the predator. Consider our nation's deer population: deprived of natural predators like wolves, herds of deer often grow to sizes that their habitats cannot support, strip the area of its vegetation, and eventually face mass death by starvation or disease.

Although a stable ecosystem is always changing in response to fluctuations, some changes can affect an ecosystem beyond its normal capacity to recover; these changes are known as stresses or disturbances. Some stresses, such as major fires, floods, droughts, hurricanes, and volcanic eruptions tend to drastically alter community structure. In these cases, it may take a long time before the community of organisms present at the time of the catastrophe again occupies that habitat. For example, Hurricane Hugo's assault on South Carolina's coast in September of 1989 had a devastating effect on the Francis Marian National Forest. Over 100,000 acres of the forest were leveled. Living in the forest were approximately 500 breeding pairs of the endangered red-cockaded woodpecker. Since the woodpecker prefers to nest in cavities of trees—trees that are no longer available—the species may not survive during the time that it takes the forest to recover. More recently, the devastation wreaked by Hurricane Andrew on South Florida in August of 1992 may affect the Florida Everglades ecosystem and its resident biota in ways yet unknown.

Culturally induced changes, such as those resulting from agriculture, urbanization, and pollution, can also disturb ecosystems drastically. One type of ecosystem that has suffered from an especially great number of culturally induced stresses is the wetland. A wetland is an area such as a marsh or swamp which is covered by water for at least part of the year. A stable wetland ecosystem provides many functions that are essential to humans as well as to nearby plant and animal communities; such areas can reduce damage from floods, filter and replenish groundwater supplies, and diminish erosion. Unfortunately, until recent years, most people were ignorant of the invaluable ecosystem services provided by wetlands; as a consequence, many of these important areas have been drained for agriculture or development.

South Florida's Everglades is an excellent example of what happens to a balanced ecosystem when human activities disturb the relationships that enable it to maintain its stability. The attractive climate of South Florida has given rise to increased population and urbanization. Growing demand for flood protection and agricultural land has led to drastic alterations in the size and natural processes of the Everglades, once a vital, interconnected ecosystem including wetlands, the keys of the Florida bay, subtropical forests, and other unique habitats. If this ecosystem, the species that inhabit it, and the valuable functions it performs are to maintain their stability, we must achieve a more equitable balance between human needs and the requirements of this complex natural system.

PHYSICAL BOUNDARIES: ESSENTIAL
FUNCTIONS PERFORMED BY THE EVERGLADES

The Everglades is a region of contrasts. It is entirely within the continental United States, yet it is as foreign to most U.S. citizens as a pristine tropical rain forest. For most people, it is at once mysterious, forbidding, and awe-inspiring. Although a watery world, the Everglades is not open water; it straddles the ocean but holds it at bay. Home to exotic animals—the Florida panther, the alligator, the snowy egret—the Everglades, like all ecosystems, is nonetheless defined by its plant life. It is a "river of grass," so very different from Florida's crowded beaches, amusement parks, and raucous nightlife that it seems a world unto itself, removed and set apart from the rest of the state. Unfortunately, nothing could be further from the truth.

The Everglades ecosystem stretches south from Lake Okeechobee across most of southeast Florida, encompassing the Florida Bay and coastal areas as well. One million four hundred thousand acres of this area was set aside as Everglades National Park in 1947 and remains the second largest national park in the 48 contiguous states. Ninety percent of the park has also been designated as a wilderness area. Despite the protection afforded by park and Wilderness designation, human demands and activities have severely stressed the intricate physical processes of the ecosystem and thus endangered its ability to maintain stability.

The Everglades is the southernmost part of a drainage basin that begins much farther north, with the numerous lakes south of Orlando in central Florida. These lakes drain southward into the Kissimmee River, which once meandered for 98 miles through thousands of acres of marshland. The river has been replaced by a 52-mile canal that quickly carries the water into Lake Okeechobee. Before the creation of a vast network of canals and dikes to control water in the Everglades region, Lake Okeechobee covered 730 square miles, although it averaged a depth of only 12 feet. Florida's heavy summer rains would periodically flood the lake, and water would spill over the lake's southern rim in vast sheets six to ten inches deep and forty to fifty miles wide. The sheets of water moved slowly—around 100 feet per day—through sawgrass, cypress swamps, and other vegetation until they emptied into the Gulf of Mexico and the Florida Bay. It was this vast, grassy wetland that inspired the area's original Native American inhabitants to give it the name of Pa-Hay-Okee, or grassy waters. During the dry winter period, the water south of the lake gradually dried up, leaving only shallow pools. Today, water in the Everglades no longer follows this seasonal cycle: Lake Okeechobee's waters flow into a series of canals that cut through the rich agricultural land produced from drained wetlands.

Ironically, this vast wetland ecosystem that has been so dramatically altered by the construction of canals and dikes, the draining of wetland areas, and the growth of developed urban areas once performed—and to an extent still performs—a number of functions that not only maintained its own stability but benefited area residents as well. Because the Everglades could absorb and store huge quantities of floodwater, it reduced damage from South Florida's frequent floods; the area's extensive vegetation slowed down floodwater currents as well. The Everglade's numerous plant species also helped to absorb energy from storms, waves, and currents, and thus decreased erosion along banks and the shoreline.

The Everglades has also played an important role in purifying and replenishing groundwater. The Florida peninsula is made of porous limestone deposited during the Pleistocene era, and small pores and channels throughout the limestone hold groundwater in large zones known as aquifers. An aquifer is not a series of underground lakes or moving rivers. Rather, it encompasses the total area of the individual pores and spaces that hold water. Most aquifers are replenished very slowly by water percolating down through the overlying ground. The Biscayne Aquifer, the source for drinking water for more than 3 million people along the Palm Beach-Miami strip of eastern Florida, was originally replenished almost entirely by rainfall and natural sheet flow in the Everglades.

The Everglades ecosystem not only recharged the aquifers beneath it, but also filtered the water that moved through its system. Before the 1930s, all areas of low elevation in South Florida were covered with water at least part of the year. Vegetation in such wetlands helps to purify water because it encourages the settling of suspended sediments, which have often combined with toxic substances or nutrients that could cause eutrophication (see description of eutrophication in Chapter 1). In addition, wetland plants take up and recycle nutrients and other potentially harmful substances such as iron.

Today, however, recharge of south Florida's aquifers depends in large part on the canal system. Because much of the water which reaches the aquifer is no longer naturally filtered by the chemical and biological processes of a wetland, water quality suffers. In addition, the draining of wetlands and increased demand on the area's groundwater have lowered the level of the aquifer. Because there is less fresh water to hold seawater back from entering the aquifer, it is now beginning to intrude.

The alteration of the Everglades' natural water cycle has also had an effect on the region's climate. The mean annual rainfall in South Florida is 60 inches; however, this may vary from

35 inches in dry years to 120 inches in wet years. This great variation is due to South Florida's distinctive rainfall cycles: periods of increasingly drier years followed by increasingly wetter years. The amount of water that evaporates from the Everglades' shallow waters is considerable, and still more water is lost through the transpiration of the area's abundant vegetation. As a result, the moisture from the rising wetlands air accounts for about 35 percent of the area's total rainfall. As the wetlands near the coast are "reclaimed" for urban development, the reduced amount of moisture and resulting increase in the heat of the ground mean that less rain falls to recharge the aquifers.

Even the fertile soil that made the wetlands attractive to farmers is endangered by human activities in the Everglades. The organic muck of the wetland was formed by decaying vegetation over thousands of years; unless it stays moist, it oxidizes, or chemically combines with oxygen. Once the soil is oxidized, its nutrients can easily wash away. When water is scarce, soil loss occurs. The most extensive soil loss has occurred south of Lake Okeechobee in the Everglades Agricultural Area, which produces about $700 million in farm products annually. Because farming requires the land to be drained, the loss of muck soil to oxidation occurs at the rate of one inch per year. The dried peat of the drained land is also extremely susceptible to fires and has increasingly been a problem throughout South Florida. Fires burn beneath the surface of the ground and are difficult to put out.

BIOLOGICAL BOUNDARIES: HABITATS AND SPECIES OF THE EVERGLADES

The Everglades' subtropical climate fosters a diverse blend of tropical and temperate species: fresh and salt water fish, many varieties of reptiles, amphibians, and mammals, over 200 species of birds, and over 1000 plant species. These plants and animals depend on the stability of the Everglades ecosystem—and especially on its seasonal water patterns—to maintain their own life cycles. As the ecosystem's stability is disrupted by human activities, though, many of these species have suffered the consequences.

The diversity of species that occupy the Everglades depends on a number of important habitats. The habitats most regularly covered with water are the sawgrass prairies and marshes, which remain wet for 5-10 months each year. The inland marshes include broad shallow rivers, small scattered ponds, and deeper holes dug by alligators. These marshes, dominated by Muhly grass and a grasslike sedge called sawgrass, are important feeding, mating, and nesting grounds for many migratory waterfowl and wading birds such as herons and egrets. Coastal marshes provide spawning and nursery grounds for fish and shellfish as well as a supply of nutrients to offshore waters. The Everglades' coastal marsh area includes one of the largest mangrove forests in the world. Many of the keys, or tiny islands, of the Florida Bay are protected refuges for nesting birds.

Drier than the marsh and bay areas, wax myrtle thickets serve as nesting and roosting grounds for many wading birds and are also important habitats for white tailed deer, bobcats, alligators, and the Florida indigo snake. Tropical hardwood hammock forests are jungle-like groves elevated roughly three feet above surrounding terrain. Covered with water only one month or less per year, these forests support a very rich and diverse biotic community: gumbo limbo trees, mahogany trees, royal palms, and saw palmettos create a home environment for such species as the endangered Florida panther, the gopher tortoise, and the colorful Liguus tree snail. Another distinct habitat is the pinelands, an unusually diverse pine forest containing roughly 200 plant species, of which approximately 30 are found only in this area.

Species that dwell in these habitats often rely on one another for feeding or other needs. For example, inhabitants of hardwood hammocks gain fire protection and a food supply from surrounding marshes, and hammocks are sources of shelter during floods for the marsh species. Alligators have a habit of digging deep ponds in otherwise shallow marsh waters; when water levels drop during the dry winter season, these ponds provide habitats for fish, which serve as a food source for different species of wading birds. The alligators in turn prey on the wading birds. Such relationships contribute to the wetland ecosystem's continued stability.

SOCIAL BOUNDARIES: THE HUMAN POPULATION AND THE EVERGLADES

Although the Everglades ecosystem has provided many benefits to the people who live nearby, only recently have we begun to realize how severely many human activities have stressed the ecosystem's stability, and thus its ability to provide those benefits.

The fertile, peaty muck soil of the Everglades has long been viewed as desirable farmland, but farmers have had to drained it of its water and clear it of its native plants before they could use it to grow crops or raise livestock. Despite the effort required, vast stretches of the area have been "reclaimed" for agriculture and now produce millions of dollars in sugar cane, vegetables, and cattle each year.

Because of Florida's mild climate and proximity to the ocean, huge population centers have grown up on the state's eastern coastal ridge. As this population increases, demand for more space pushes back the wetlands. Before 1900, 75 percent of South Florida was wetland. Today about 35 percent of that 75 percent has been converted to agricultural or urban development. In addition, the growing populations in these areas combine with the region's farms to demand more and more water for drinking and irrigation from the Everglades and its underlying aquifers.

The Everglades ecosystem also holds great recreational, aesthetic, and scientific value. Tourism is one of Florida's most important industries, and hunting, fishing, and birdwatching are only a few of the activities offered in the wetlands. Sportfishing on Lake Okeechobee, especially famous for its bass population, generates as much as $40 million a year. In addition, the diverse and productive ecosystem provides many opportunities for research, which is especially important because of the potentially serious environmental problems now facing the area. For example, the South Florida Research Center is currently conducting studies in hydrology, wildlife ecology, marine ecology, and plant ecology in order to determine how changes outside the park affect the ecosystem and how it might best be protected from such disturbances. Information gleaned from such studies can benefit not only the Everglades but other threatened wetlands ecosystems as well.

LOOKING BACK: THE HISTORY OF THE FLORIDA EVERGLADES

It was not until after the Civil War that human activities began to seriously affect the stability of the Everglades ecosystem. In 1881, a young businessman from Philadelphia named Hamilton Disston purchased 4 million acres in the Everglades. His goal was to drain land north of Lake Okeechobee for cultivation and build a navigable waterway between the lake and the Gulf of Mexico. In slightly more than a decade, he had drained 50,000 acres for the cultivation of rice, sugar, and peaches. Although Disston eventually went bankrupt, numerous other parties—both private interests and the government—rushed to follow his lead in draining the area's wetlands and controlling its water. By the mid-1920s, 440 miles of canals, 47 miles of levees, or raised banks, and 16 locks and dams had been constructed.

The network of canals, levees, and dams in the Everglades was created not only to improve navigation of inland channels and to convert marshland to agricultural uses, but also to offer the area's population increased flood protection. Although the wetlands of the Everglades naturally slowed and contained floodwaters, many cities, farms, and other enterprises had been built in floodplains that were frequently flooded because of their low altitude and proximity to rivers, lakes, and marshes. After hurricanes in 1926 and 1928 which left thousands of people dead, demands for better flood protection generated increased construction in the Everglades; as a result, the southern rim of Lake Okeechobee was replaced by the Herbert Hoover Dike, which completely halted the normal sheet flow. A record drought in 1944-1945, followed by hurricanes in 1947 and 1948 which caused over $150 million in damage, prompted further control efforts. As a result of subsequent years of construction, 1350 square miles south of Lake Okeechobee have been converted into a series of water conservation areas: impoundments where water levels and movement are carefully regulated through a network of almost 1500 miles of levees and canals, dams, tide and flood gates, and water pumping stations. This system, which controls water conditions throughout most of the southern Everglades, is managed by the U.S. Army Corps of Engineers and the South Florida Water Management District (SFWMD). In general, the system keeps water levels in Lake Okeechobee low by draining billions of gallons of freshwater into the Atlantic and the Gulf of Mexico.

North of Lake Okeechobee, the Kissimmee River has been drastically altered as well. The Kissimmee River was originally the water source for perhaps as many as 200,000 acres of river marsh and associated wetlands. Millions of migratory and wading birds depended on this wetland habitat. However, area residents wanted to expand the recreational and navigational uses of the river and to create better flood control for the urban area south of Orlando and the farmers and ranchers of the lower basin. As a result, the Army Corps of Engineers undertook a project to replace the winding river with a straight canal 52 miles long, 200 feet wide, and 30 feet deep. Completed in 1971, the waterway was named Canal 38.

With the transformation of the Kissimmee into a canal, water no longer took eleven days to reach Lake Okeechobee; it rushed through in two. Nitrogen and phosphorus runoff from the newly created pastures and farmland along the canal and the waste water from the Orlando area moved quickly downstream, allowing few of the pollutants to be filtered out. The old oxbows of the meandering river slowly silted up and disappeared. As the wetlands along the river basin disappeared, so did the wildlife. The bald eagle population decreased greatly and ducks and coots nearly disappeared.

As a result of such intensive human interference in the name of better flood control, improved navigation, and expanded farmland, only about half of the area's original 4 million acres of wetlands remain today. The National Park contains about 500,000 of them.

The Effects of a Disrupted Water Cycle, Increased Agriculture, and Exotic Species

Perhaps one of the most telling signs of the Everglades ecosystem's threatened stability is the dramatic decline in many of the unique species that depend on it. The number of nesting wading birds in the Everglades has dropped from 300,000 in the 1930s to 15,000 today. Although 250,000 pairs of white ibises made their home on the northern boundary of the Everglades in the 1930s, today no large nesting colonies remain on the Everglades National Park mainland. The past 25 years have seen an 80 percent drop in breeding pairs of the endangered wood storks. In most years since 1980, fewer than 1000 pairs of great egrets and 500 pairs of snowy egrets have nested in mainland colonies. Officially listed as endangered are the Florida panther, the West Indian manatee, the American crocodile, the Everglades snail kite, the southern bald eagle, the peregrine falcon, the "Cape Sable" seaside sparrow, the hawksbill turtle, the green turtle, and the Atlantic Ridley turtle. The loggerhead turtle, the indigo snake, and the American Alligator are classified as threatened.

Although complex factors lie behind the decline of these and other species, a great deal of the blame can be placed on the consequences of massive growth of the area's human population: construction of a system of canals and dams to control the area's water cycle, the conversion of natural wetlands to agricultural fields, and the introduction of exotic species.

As we have seen, water is the lifeblood of South Florida's Everglades. However, human attempts to control the water cycle have disrupted the natural pattern of water flow through the ecosystem, drastically altering many factors contributing to its stability. In addition to interfering with ecosystem functions such as groundwater recharge and purification, the disruption of the water cycle has interfered with the life cycles and population balance of many Everglades species.

Although the Everglades originally experienced a yearly cycle of wet summers and dry winters, the SFWMD has filled and withdrawn water according to a schedule based purely on human needs. When floods or heavy rains threaten urban or agricultural areas, managers use the Everglades as a reservoir to absorb excess water, and they draw water from the Everglades when it is needed for irrigation or drinking. Unfortunately, this process has caused the Everglades to experience extremes of wetness and dryness much greater than those caused by the natural water cycle. These extremes can be disastrous for the species that live there.

From 1980 to 1982, South Florida suffered its worst drought in history. With the increase in dry land in the water conservation areas below Lake Okeechobee, the white-tailed deer population boomed to an estimated 5500. In the spring of 1982 heavy rains began to raise the water levels. Flood water was channeled into the conservation areas until only a few hundred acres of dry land were left. Under these conditions the deer population could not survive. The Game and Fresh Water Fish Commission called for an emergency out-of-season deer hunt in July 1982, to thin the herd and increase the chances of survival for the remaining deer. A two-day hunt was held and 723 deer were killed. The rest of the hunt was cancelled in favor of a rescue attempt. Nineteen deer were captured in a two-day period; only six survived. The Commission estimated that 1200 deer may have died in one area as a result of high water in 1982.

Occasionally during the rainy season the Water Conservation Areas cannot contain all of the water drained from surrounding land. In such cases, some of the excess water is pumped into the Everglades National Park. This often floods the park at the worst possible time, when it already has enough water from rainfall and when the alligators and wading birds are nesting. In three of the four years before 1983, a significant percentage of the area's alligator eggs were lost to artificial flooding, and wood storks have had successful nestings in only three of the past 18 years. As a result of torrential rains in 1983, floodgates into Everglades National Park released so much water that wading bird nesting was destroyed. Sudden extremes of water can be disastrous outside the National Park as well. In August 1988, when 12 inches of rain flooded farmlands in the area, the SFWMD directed the floodwaters into a sound adjacent to the park. This influx effectively destroyed the sound's population of breeding shrimp and fish, both important food sources for humans and wading birds.

Another major stress to the Everglades' stability originates from the farmlands that surround it. The Everglades Agricultural Region south of the lake contains 700,000 acres, over two-thirds of which is used to grow sugar cane. Water running off from these fields into the lake contains between ten and twenty times the normal level of phosphorus and nitrogen due to the application of fertilizers. In addition, waste from the roughly 45,000 dairy cattle raised north of the lake contributes phosphorus-laden runoff to the lake. This heavy load of nutrients has stimulated the lake's cultural eutrophication, a heavy growth of algal blooms that deplete the waters' oxygen and thus threaten the lives of many of its species.

Even though the SFWMD responded to this problem in 1979 by ceasing to pump run-off from agricultural fields into the lake except during drought or flood situations, in 1986 algal blooms still covered 120 square miles of the lake. Further efforts to diminish the algal blooms by diverting runoff to the Water Conservation Areas simply resulted in new problems: the heavy nutrient load caused shifts in the natural mix of vegetation. On as many as 30,000 acres, cattails have overrun the area's dominant sawgrass, blocking light from other species, including the periphytic algae that are the base of the Everglades food web.

Another threat to the Everglades ecosystem is the encroachment of exotic or nonnative plant species. When exotic species are introduced to an area, they often have no natural competitors of predators. As a result, they sometimes overrun an ecosystem, altering its normal pattern of succession, or sequential changes in species composition. In damaged and drained lands it is especially easy for exotics to move in and dominate. Although a variety of exotic species, including the Brazilian pepper, the Australian pine, and the Melaleuca threaten other species in the Everglades, the problem of the Melaleuca serves to illustrate the danger of exotic species.

In 1906, a forester of the Department of Agriculture introduced the eucalyptus-like Melaleuca because he thought it would attract commercial woodcutters. But inside its soft, thick bark the hardwood interior is runny with water and difficult to saw, which makes it unattractive to the timber industry. The tree uses three times as much water as other trees and has taken over about 60 square miles of wetland. Its leaves are filled with eucalyptol, an oily, flammable substance that smokes and sparks when on fire. Fire does not kill the tree, however; it is protected by its insulating bark.

Within Everglades National Park there are millions of Melaleuca over hundreds of square miles. Although park employees apply herbicides to remove "outlyers" and try to keep the trees out of the pinelands, no effective solution for this problem has been found so far.

Progress Toward Safeguarding the Everglades Ecosystem's Stability

Although humankind's influence on the Everglades has primarily been one of exploitation and disturbance, we are gradually realizing that even such a vast, rich resource cannot maintain stability under unremitting stress. This realization has been accompanied by efforts to restore and protect this unusual ecosystem.

By 1983, biologists concerned about the effects of altering the ecosystem's water cycle urged the SFWMD to adopt a water delivery pattern designed to mimic seasonal rainfall rather to respond exclusively to human needs. Use of a sophisticated computer model has allowed water managers to adapt the timing and distribution of water releases in the Everglades to more closely resemble the area's natural water cycle. However, water levels are still too low because the new water delivery schedule was based on the rainfall patterns of a 15 year period that included three severe droughts. Most recently, the federal government has initiated a $500,000 study to evaluate ways to further improve the park's water supply.

Another important improvement in human management of the Everglades has been the creation of a buffer zone around its perimeters to help protect it from the effects of human activities. In 1983, 55,000 acres were added around the boundaries of the park. In 1989, the Everglades Expansion Act led Congress to approve an expenditure of $15 million in order to appropriate 107,000 additional acres. The acreage includes half of the Shark River Slough, an area that once played an important role in delivering water from Lake Okeechobee to the Everglades. Although the park was originally supposed to include the entire slough, compromises resulted in half the area being sold to developers. Fortunately, little development occurred in the slough, but a series of canals and dikes built for flood protection must now be removed by the Army Corps of Engineers as part of a plan to recreate original water patterns on the additional 107,000 acres.

The Kissimmee River has been the site of restoration as well. The state of Florida appeared to regret its alteration of the Kissimmee River as early as 1973, when it passed the Kissimmee Restoration Act pledging to return the river to its original role in Everglades ecosystem. It was not until 1984 , though, that the SFWMD began experiments designed to determine if the Kissimmee could be restored: they began by reflooding old oxbows and marshlands along a short stretch of the river, and then spent three years manipulating water levels to mimic seasonal water variations. Although costs for deconstructing the canal and purchasing the 50,000 acres of floodplain now being used for pasture could amount to as much as $300 million, the SFWMD has concluded that the restoration of the Kissimmee is possible.

Apparently concerned with the effects of nutrient loading to Lake Okeechobee and the water conservation areas to its south, the federal government actually filed a lawsuit against the SFWMD in 1990. The suit claims that, by failing to enforce state water quality standards, the SFWMD has put at risk both the Everglades National Park and the federally owned Loxahatchee National Wildlife Refuge, which is in the Everglades watershed. In response to this unexpected intervention by the federal government, in 1991 the SFWMD adopted the Surface Water Improvement and Management Plan. According to the Plan, phosphorus loading from agricultural fields will be reduced by 75 percent by converting 18,000 acres of sugar cane and sod farms into wetlands. In addition to reducing the amount of fertilizer absorbed by runoff, this conversion would increase the wetland area available to absorb excess nutrients. However, critics of the plan claim a full 93,000 acres of wetlands would have to be restored to achieve the plan's goal.

LOOKING AHEAD: THE FUTURE OF THE EVERGLADES ECOSYSTEM

In recent years, the federal government and the SFWMD have taken various steps to restore the Everglades ecosystem to a healthier, more stable condition. However, the history of human interaction with the Everglades shows that the involvement of key individuals as well as grass roots support has also played an important role in increasing public awareness of the Everglades' plight and in spurring action to protect the ecosystem.

Perhaps the most well known of all individuals working for the Everglades is Marjory Stoneman Douglas. An environmental activist who has lived in Florida since 1915, Douglas wrote a poetic account of the mystery and beauty of the Everglades in her 1947 book, *The Everglades: River of Grass*. In 1970, Marjory Douglas founded Friends of the Everglades, an organization that has fought to preserve the Everglades ever since. Through private lobbying and public hearings, the group has accomplished many important goals, which include blocking the construction of a jetport in the East Everglades and forcing two drainage canals to be plugged.

Another important figure in the fight to protect the Everglades is Robert Graham, who served as governor of Florida from 1978 to 1986. During his term, he organized Save Our Everglades, a movement that has proven very effective in the effort to restore this vital ecosystem. One of Save Our Everglades' most important projects has been its support of the plan to restore the Kissimmee River. Although the restoration was begun in 1984, a great deal of work still needs to be accomplished: the water must be completely rerouted into its former course, land must be acquired, dikes removed, and approximately 28,000 acres of wetlands restored. Most important, water level fluctuation must be returned to its natural cycle.

Although a few area residents object to the restoration, hydrologists from SFWMD say that a reflooding of the Kissimmee River valley would be better for everyone except the property owners along the river, who will be losing flood protection. Towns near the upper lakes will not be threatened, because the canal never provided the expected flood protection in that area. In addition, restoration will reduce the need for emergency releases of water—sudden dumpings that cause damage in the coastal estuaries by drowning oyster and clam beds and by depositing silt.

Despite the improvements already underway, efforts to safeguard the stability of the Everglades ecosystem must continue. The Army Corps of Engineers should consider restoring even more areas where canal and dam systems have disrupted natural water patterns. One such area is the C-111 Canal, which diverts water from Taylor Slough (southeast of the park). Use of this canal for flood control has forced fresh water into salt water habitats such as Manatee Bay and Florida Bay, with lethal results for resident fish and crustacean species. In addition, federal and state agencies responsible for managing the area should strive for better cooperation; together, they must decide on a course of action which takes into account not only human needs, but also the long-term viability of the Everglades' habitats and species. Continued research is necessary to determine the effects of human activities on these habitats and species as well as the best ways to protect and restore them.

Although past and future efforts to institute a more natural water cycle, replace farmland with wetlands, and otherwise safeguard the Everglades' stability are essential ones, they will have little effect if no attention is given to the area's booming human population. The population of Florida has increased in the last 40 years from 2 million to over 10 million. Currently, Florida residents use an average of 200 gallons of water per person per day—twice the national average. Area farmers pay nothing for irrigation water, and so have no incentive to conserve. In the future, the balance between human needs, water, and wetlands will become even more strained: Florida's population is expected to double from its 1975 level by the year 2000. Serious study, thought, and action must be given to the best ways to meet the needs of this population while one of its most precious resources is preserved.

SUGGESTED READING

Belleville, Bill. "Turnaround for the Everglades." *Oceans*. July/August 1986. 16-21.

Bolgiano, Chris. "Of Panthers and Prejudice." *Buzzworm: The Environmental Journal*. May/June 1991. 47-51.

Boucher, Norman. "Smart as Gods: Can We Put the Everglades Back Together Again?" *Wilderness*. Winter 1991. 10-21.

Carr, Archie Fairly. *The Everglades*. New York: Time-Life Books, 1973.

Caulfield, Patricia. *Everglades: Selections from the Writings of Peter Matthiessen*. San Francisco: Sierra Club, 1970.

De Golia, Jack. *Everglades; The Story Behind the Scenery*. Las Vegas: KC Publications, 1978.

Douglas, Marjory Stoneman. *The Everglades: River of Grass*. Revised edition. Miami, FL: Banyan Books, 1978.

Duplaix, Nicole. "South Florida Water: Paying the Price." *National Geographic*. July 1990. 89-114.

Flowers, Charles. "Starting Over in the Everglades." *National Wildlife*. April/May 1985. 54-63.

Graham, Frank. "The 'Sewer Ditch' Undone." *Audubon*. March 1987. 114-15.

Hansen, Kevin. "South Florida's Water Dilemma: A Trickle of Hope for the Everglades." *Environment*. June 1984. 14+.

Johnson, Lamar. *Beyond the Fourth Generation*. Gainesville, FL: University of Florida Press, 1974.

Kahn, Jeffery. "Restoring the Everglades." *Sierra*. September/October 1986. 38-43.

Lewin, Ted. *World within a World—Everglades*. New York: Dodd, Mead, 1976.

Light, Stephen S., John R. Wodraska, and Sabina, Joe. "The Southern Everglades: The Evolution of Water Management." *Phi Kappa Phi Journal*. Winter 1989. 11-14.

McMullen, James P. *Cry of the Panther: Quest of a Species*. Englewood, FL: Pineapple Press, 1984.

Mitchell, John G. "The Perils of the Everglades." *Wilderness*. Winter 1986. 12+.

Mitsch, William J., and James G. Gosselink. *Wetlands*. New York: John Wiley, 1992.

Monks, Vicki. "Engineering the Everglades." *National Parks*. September/October 1990. 32+.

Pardue, Leonard, et al. *Who Knows the Rain? Nature and Origin of Rainfall in South Florida*. Coconut Grove, FL: Friends of the Everglades, 1982.

Postel, Sandra. "The Not-For-Everglades?" *World Watch*. March/April 1991. 34.

Schneider, William J., and James H. Hartwell. "Troubled Waters of the Everglades." *Natural History*. November 1984. 46-57.

Sobel, Dava. "Marjory Stoneman Douglas: Still Fighting the Good Fight for the Everglades." *Audubon*. July/August 1991. 30-39.

Tebeau, Charlton W. *Man in the Everglades; 2000 Years of Human History in the Everglades National Park*. Second Revised Edition. Coral Gables, FL: University of Miami Press, 1968.

Toops, Connie M. *Everglades*. Stillwater, MN: Voyageur Press, 1989.

Wetlands: Their Use and Regulation. Washington, DC: U.S. Congress, Office of Technology Assessment, OTA-O-206, March 1984.

Williams, Michael, ed. *Wetlands: A Threatened Landscape*. Cambridge, MA: B. Blackwell, 1990.

CHAPTER *4*

ECOSYSTEM DEGRADATION: THE BLUE, THE GRAY, AND THE HUMPBACK WHALES

The past three chapters have explored the structure, function, and dynamic equilibrium of particular ecosystems. However, our discussions of Lake Erie, the Chesapeake Bay, and the Everglades also included the effects of human activities on these ecosystems. In many cases, stresses and disturbances caused by human activities resulted in serious problems, including cultural eutrophication, polluted or diminished water supplies, drastic reductions in populations of certain species, and loss of important habitats such as wetlands. In this chapter, we will focus specifically on the ways that our society contributes to the degradation of ecosystems.

A natural system becomes degraded when human activities alter environmental conditions in such a way that they exceed the range of tolerances for one or more organisms in the biotic community. For example, when increased amounts of nitrogen and phosphorus from agricultural run-off cause algal blooms to spread across a lake, the decomposition of the algae eventually lowers the oxygen content of the lake. When the oxygen content falls below a certain concentration, some fish species can no longer survive—the oxygen level has become lower than they can tolerate. When an ecosystem is degraded, it loses some capacity to support the diversity of life forms that are best suited to its particular physical environment.

Ecosystem degradation can take many complex forms, but most can be classified into one of five broad categories that we call the Five D's of natural system degradation: Desertification, Deforestation, Destruction, Damage, and Disruption.

Desertification and deforestation, which can drastically alter the appearance of an area, are two fairly obvious ways human activities degrade natural systems. Desertification is the impoverishment of land due to overgrazing, overcultivation, deforestation, or poor irrigation practices. It degrades natural systems by spreading, expanding, or creating desert-like conditions in areas where such conditions do not naturally occur. Desertification primarily affects agricultural lands and grasslands, and is increasing at an alarming rate throughout the world. Deforestation is the cutting down and clearing away of forests; this practice not only eliminates habitat but also contributes to increased erosion because the trees' roots no longer hold the soil in place.

Destruction, which completely replaces natural systems with human systems, is in some ways less obvious than desertification and deforestation because it leaves no trace of the original ecosystem to reminded observers that it has been degraded. Urbanization, transportation, and agriculture are the major causes of destruction. As the human population increases, we continue to destroy the habitat of other species at an accelerating rate.

Damage and disruption are the least obvious forms of degradation because ecosystems continue to exist and may even fail to show visible or quantifiable changes until some time after the damage or disruption has occurred. Examples of damage include pollution, siltation, and the encroachment of saltwater into freshwater wetlands because of channeling or the removal of groundwater. All types of damage violate the integrity of ecosystems, making it more difficult for their species to survive.

Although damage affects an ecosystem's species indirectly, disruptions have a direct impact on an ecosystem's biota. Disruptions are human activities that directly lead to rapid changes in the species composition of a community. Changes in species composition have been linked to the use of persistent pesticides; the introduction of new species into an area; construction projects such as dams or highways; and overexploitation of resources. When species disappear or when drastic population changes occur within a community, other species, and the ecosystem in general, are likely to be affected as well.

A dramatic example of the way the extinction of a single species can affect the rest of the ecosystem occurred in North Borneo in 1955. In response to a high incidence of malaria in the country, the World Health Organization sprayed dieldrin (similar to the pesticide DDT) to kill the mosquitos that carried the protozoan parasite that causes the disease. Although the dieldrin killed most of the mosquitoes and nearly eliminated malaria from the island, it also killed many other insects, including flies, wasps, and cockroaches. Small lizards that fed on the dead insects died as well. Soon, cats began dying after feeding on the dead lizards. Because cats had been the main predators of rats in the area, before long, the rodent population had increased significantly. Rats began to overrun the villages, and the people faced a new threat: sylvatic plague, caused by a bacterium carried by fleas on the rats. The situation was brought under control when the British Royal Air Force, on instructions from the WHO, parachuted cats into Borneo.

As we can see, eliminating a species from an ecosystem can seriously disrupt that system. This chapter will examine the possibility that several species of whales—irreplaceable members of the ocean ecosystem— may soon become extinct because of human activity. Although we know little about the whale's role in that ecosystem, scientists speculate that whale remains may play an important part in the life cycles of organisms that dwell on the ocean's floors. In addition, the whale's tremendous size, beauty, and other unique qualities have led many of us to see this striking animal as a symbol of the plight of all species that are endangered by human activities. We know that whales are ancient inhabitants of earth, having first lived on land and then evolved to adapt to life in the ocean approximately 60 million years ago, long before the appearance of humans. They are capable of communication and are potentially intelligent, possessing a large and complex brain.

In addition to calling attention to the way the extinction of species like the whale can disrupt natural systems, this chapter also illustrates the potential effects of an ecosystem's damage on its species. Pollution, global warming, and the results of other human activities may have already damaged the whale's habitat irreparably, and we have yet to learn the complex ways such damage may have affected whale species.

In this chapter, we will focus on three species of whale: the blue, the gray, and the humpback. All were hunted extensively in the past and, although they are now protected by international law, their populations have yet to significantly recover. Before the advent of whaling, the total population of blue whales is estimated to have been 200,000 individuals; pre-whaling populations of gray and humpback whales are believed to have been approximately 24,000 and 125,000 individuals, respectively. In 1988, *National Geographic* reported a total global population of just 5-10,000 blue whales, 21,000 gray whales, and over 12,000 humpback whales. In addition, these three species represent a broad range of sizes, geographic range, and habits of feeding, breeding, and communicating. By taking a close look at the blue, the gray, and the humpback whale, we can gain a greater appreciation for the value and plight of all species of whales as well as the importance of every component of an ecosystem.

PHYSICAL BOUNDARIES: WHERE THE BLUE, GRAY, AND HUMPBACK WHALES CAN BE FOUND

The blue, the gray, and the humpback whale all follow somewhat predictable seasonal patterns of migration. Because whales are homeothermic (warm-blooded) mammals, they feed in colder waters only during the summer months, migrating to warmer waters with the coming of autumn. There, they breed and give birth to their young.

Because blue and humpback whales are found throughout the world's oceans, the migratory routes they follow depend on the stock to which they belong. A stock is a grouping of whales of the same species which are found in the same general location. For example, the Southern Ocean and North Pacific groups are two stocks of humpbacks. A third humpback stock resides in North Atlantic waters, wintering in the Caribbean and feeding off Cape Cod and the Gulf of Maine in summer. Stocks are composed of pods, or groups, of whales that migrate together. Humpback pods typically include three or four individuals, whereas a blue pod consists of two or three whales.

The gray whale, also known as the California gray whale, can be found only in the North Pacific. Each autumn, gray whales leave their summer feeding grounds in the Bering and Chukchi Seas and migrate toward Baja California in pods of up to six members. There, in the warm waters of sheltered bays and lagoons they court and mate, and cows give birth to the young who were conceived the previous breeding season. From late January to April, the whales again migrate, heading northward to their Arctic feeding grounds. They feed in colder water during the summer because of the incredible productivity of the Arctic waters during the long summer days.

The routes followed by migrating whales are important for several reasons. First, their strict migrational patterns traditionally enabled hunters to easily determine the times and places certain whales could be found and harvested. And, because these routes cover such a vast expanse of territory, whales are a true global resource. It is difficult to manage a whale stock; although one country may protect the stock within its national waters, it cannot protect those whales after they have migrated elsewhere. Clearly, managing whales means protecting them throughout their range. International protection against hunters and other factors such as pollution is essential if whale populations are to be restored to healthy levels.

BIOLOGICAL BOUNDARIES: HOW THE BLUE, GRAY AND HUMPBACK WHALES FEED, BREED, AND COMMUNICATE

Based on fossil evidence, most scientists believe that whales were once terrestrial mammals who ventured into water in search of food or sanctuary. As eons passed, the whale changed significantly in adaptation to its new aquatic environment. The tail broadened into flukes, allowing the animal to propel itself through the water. The development of a thick, smooth blanket of blubber helped the whale to maintain its body temperature even in cold polar waters. Nostrils at the top of the head enabled the whale to breathe by surfacing. The buoyancy of the watery world brought another significant change: freed from the restrictions of the land, the body of the whale became enormous.

Although all whales share these common characteristics, they also evolved into dozens of distinct species. All whale species fall into the biological order Cetacea, which is made up of two groups: toothed whales (Odontoceti) and baleen whales (Mysticeti). Toothed whales are composed of approximately 65 species, including sperm whales, porpoises, and dolphins. All are fast-swimming hunters, using teeth and jaws to capture and grasp prey, typically fish and squid. In contrast, baleen whales have horny plates of baleen hanging down from their upper jaws. They use their baleen plates to strain food, particularly tiny crustaceans known as krill, from the water. The baleen whales are all relatively large and include six major species. In order of increasing size, these are the gray, humpback, right, bowhead, fin, and blue. In the past, baleen whales have been hunted more extensively than toothed whales because they are generally larger and slower-moving.

The blue whale takes its name from its characteristic slate blue color. One of its common names is Sulphur Bottom Whale, a reference to the yellowish film of diatoms that develops on its underside. The largest animal species on the planet, the blue whale reaches an average of 80-85 feet in length and 100 tons in weight. The largest blue whale ever recorded was 110 feet 2 1/2 inches long, over one-third the length of a football field. The gray whale is actually black. Its mottled gray appearance is caused by white parasitic barnacles which attach to the whale's skin. The average gray whale ranges between 39 and 45 feet in length and weighs 35 tons. The humpback takes it name from the way in which it dives, rounding and exposing its back and dorsal fin. Generally more acrobatic than its fellow whales, the humpback seems to enjoy performing complicated gyrations. Its flippers are flexible and very large, often one-third the size of its body. The humpback uses them for balance and propulsion. Additionally, courting pairs use their flippers to embrace and caress one another, and cows often gently pat or caress their calves with their flippers. Humpbacks weigh approximately 60 tons and reach an average of 50-55 feet in length.

These three species are also characterized by an interesting range of feeding, breeding, and communication habits.

Feeding Habits

Blue whales feed almost exclusively on krill, but occasionally feed on larger organisms, such as fish. The diet of humpbacks consists of sand lances, anchovies, sardines, and krill. Gray whales feed on krill and plankton.

Blues and humpbacks typically feed in one of several ways. They may swim slowly along with their mouths open, straining huge numbers of krill from the water. Or they may take large gulps of water, which is then expelled, leaving behind krill and other organisms. The whale then uses its tongue to scrape the food from the baleen. Humpbacks have devised another method for obtaining food. A group of seven or more individuals will dive in unison. Then, swimming toward the surface in a spiral fashion, they form a bubble net, which concentrates a frightened group of smaller fish or krill. Mouths open, the humpbacks rise through the concentrated prey.

The gray, unlike other baleen whales, feeds on benthic, or bottom-dwelling organisms. Using its short, stiff baleen plates, the gray plows through bottom sediments, sifting crustaceans, mollusks, bristle worms, and other animals.

Breeding Habits

Baleen whales prefer to breed and give birth to young in the warm and shallow waters of isolated lagoons. Before mating actually occurs, whales engage in extensive physical contact, stroking and embracing one another with their flippers. This contact alerts the male that the female is in heat, a necessity since whales lack highly developed olfactory devices. To mate, most species rise out of the water toward one another in an upright position. Gray whales, however, lie on their backs in the water and roll toward one another to make contact.

The gestation period for baleens is relatively long: 15-16 months for blues, 11-12 months for humpbacks, and 13 months for grays. Hence, on average, a baleen whale gives birth to a single calf every two years. Calves are born in shallow waters since they cannot immediately swim. The cow supports and helps her calf, bringing it to the surface to breathe. Mother whales are usually helped by other females, known as surrogate mothers or "aunts," which do not have their own calves to raise.

Blue whales are 20-27 feet long at birth and weigh approximately three tons. The milk of blue whale cows contains 35 percent fat, allowing suckling calves to add as much as 200 pounds a day. When the calves are weaned at one year, they have increased their weight ninefold, to approximately 26 tons. Newborn humpbacks and grays are approximately the same size, about 14 and 14-16 feet in length, respectively. At birth, gray calves weigh between 1500 and 2000 pounds. They grow quickly on their mothers' rich milk, which contains 53 percent fat. By the time they begin their first northward migration, gray calves are three or four feet longer and have doubled their weight to two tons.

Communication Habits

Although whales have no vocal chords, they do make a variety of sounds that are thought to be produced by the larynx and respiratory system, including the blowhole. All baleen whales produce two basic sounds, clicks and grunts. Through a process called echolocation, clicks help the whale to find its way about the environment. Grunts are thought to be a form of communication, though research thus far has been inconclusive.

Perhaps the most famous form of communication among sea mammals is the songs of the humpback whale. Humpback vocalizations are composed of about six themes, the themes of several phrases, and each phrase of two to five different sounds. Arranged in organized sequences approximately 10-15 minutes long, humpback vocalizations can be heard for thousands of miles underwater. Humpbacks sing only during the breeding season, and only the males sing, lending credence to the theory that the songs are used for mating purposes.

Although the whales do not sing while at their summer feeding grounds, they resume their singing when they return to their breeding grounds each fall.

SOCIAL BOUNDARIES: CULTURALLY PRODUCED PRESSURES ON WHALE POPULATIONS

Human social systems exert both direct and indirect pressures on whale populations. Historically, direct pressures, in the form of whale harvesting and exploitation, have been of greater significance. In the future, indirect pressures, including habitat degradation and the depletion of whales' food sources, may pose the chief threat to the great whales' survival.

Direct Pressures

Traditionally, whales were hunted to provide a variety of products, including oil, baleen, meat, and bones. Whale oil was historically one of the major products yielded by the whaling industry. Blubber was stripped from the whale and cooked, yielding oil used for margarine, soap, and cooking oil. Sperm whale oil was unsuitable for cooking purposes, but was widely used as a lubricating agent and to tan high-quality leather. Spermaceti, an oil found only in a reservoir of the sperm whale's large head, was also used to make cosmetics, crayons, candles, various polishes, and other miscellaneous items. Today, however, most products that formerly used whale oil rely on cheaper, often synthetic, substances.

Because of it strength and flexibility, baleen was widely used during the late 1800s and early 1900s in such products as umbrellas, corset stays, fishing rods, nets, and brushes. Today, the market for baleen is limited, though it is used occasionally in brushes and, in the Far East, in novelty products such as shoehorns and tea trays.

Although there is little demand for whale oil or baleen today, whales are still hunted for their meat. The Inuits of Alaska hunt whales for consumption as part of their traditional lifestyle. However, a greater pressure is placed on whale populations by countries such as Japan, Iceland, and Norway, which actively hunt whales and sell their meat. Nations that harvest whales also use the meat in the production of dog food and fertilizer, and both whale meat and bone are ground to produce livestock feed. The bones of whales are sometimes used as building materials or in an art form called scrimshaw, the etching of pictures into bone. Japan harvests more whales than any other remaining whaling nation, and it also buys a great deal of the whale meat produced by Norway and Iceland. Japan's persistent whaling operations can be traced to its cultural heritage. Whale meat was once an important element in the Japanese diet. The Japanese also used the oil, whalebone, skin, and various internal organs to produce a variety of objects, including sandals, shoelaces, and cosmetics. Today, although less than 6 percent of the population of Japan eats whale meat, a significant market exists for the meat as a delicacy.

Indirect Pressures

Although the direct exploitation of whales continues, it has slowed somewhat as other products are increasingly used in place of those obtained from whales. However, indirect pressures, which threaten the habitat and food sources of the whales, are becoming an increasing concern.

The indirect pressure that poses the greatest threat to the survival of whales is water pollution, which is certain to become of greater concern as long as the practice of dumping wastes into the ocean continues. Little is known about the effects on the ocean biota of dumping wastes, including hazardous or toxic wastes. However, if toxins build up in the food chain, they will be passed along to whale species. Other species that have ingested large amounts of toxins have eventually developed symptoms such as tumors, deformations, and infertility.

Another indirect pressure on whale populations is increasing noise pollution due to rising numbers of ocean-going vessels. Scientists speculate that the added noise may interfere with whale communication. In addition, phenomena such as acid precipitation and global warming could lead to changes in the chemical composition and temperature of the world's oceans. It is difficult to predict how such degradation of their habitat might affect whale populations. Pressure is also indirectly exerted on whale populations as humans begin to harvest krill for human consumption.

Indirect pressures are more difficult to control than direct ones for several reasons. They are widespread, occurring throughout the worlds' oceans. Also, our knowledge of the ocean and its biota, including the great whales, is incomplete. There may be many indirect pressures, caused by human activities, of which we are presently unaware.

LOOKING BACK: THE HISTORY OF HUMAN INTERACTION WITH WHALES

Although the history of human interaction with whales has been one of short-sighted exploitation for centuries, the twentieth century has seen a growing public awareness of the whale's value and plight. Many whale species are now protected from commercial harvesting, but other threats remain.

The Development of the Whaling Industry

Whales were originally hunted primarily for subsistence purposes, with local communities depending on the animals for food and fuel. The number of whales taken depended on both the needs of the community and the skill of its hunters. However, a new attitude toward whaling took hold with the development of commercial whaling by the Norse and Basque people of Europe somewhere between 800 and 1000 A.D. Rather than harvesting as much as they needed, they harvested as much as they could. Other European peoples, including the Dutch, and the Japanese followed suit around the 1600s. About this time, the North American whaling industry was born in Nantucket, Massachusetts, and quickly spread to most of the east coast. Thus, the seventeenth century ushered in the global exploitation of whales.

Until the mid-1800s, whaling was a difficult task. Whalers used small wooden boats, hand-held harpoons, and nets to pursue slow-swimming whales and those that stayed close to shore. The first species to be overexploited was the right whale, so named because it floated when killed and was thus considered the "right" whale to hunt. By the early nineteenth century, populations of right, bowhead, sperm, and humpback whales had seriously declined because of overharvesting. A similar fate befell the gray whale: in the 1840s, whalers found their way into the eastern Pacific, pursuing the gray whales in their breeding lagoons along the Baja coast and later in their northern range. By the end of the century, the population, which may have originally numbered 24,000, had fallen to only a few thousand.

The mid-1800s saw the development of improved technology for whaling, chiefly in the form of a harpoon gun mounted on a steam-powered ship. A one-ton cannon fired a massive harpoon deep into the whale's body. A fragmentation bomb contained in the nose of the harpoon then exploded, scattering shrapnel throughout the animal's internal organs. Steel barbs located along the harpoon shaft sprung outward with the force of the explosion, firmly anchoring the harpoon and an attached rope in the victim's flesh. After the dead whale had been hauled to the surface, a hollow metal tube was thrust into its lungs or abdominal cavity. Compressed air or steam injected into the whale through the tube inflated the corpse until it was buoyant enough to be towed back to the factory. This new harpoon gun allowed whalers to turn their attention to faster-swimming species.

In 1905, large numbers of fin and blue whales were discovered in the Antarctic. This discovery bolstered the faltering whaling industry, which had become the victim of its own overzealous harvesting and the resultant depletion of stocks. Whaling became even more profitable with the introduction, in 1925, of the pelagic (ocean) factory ship, which allowed whalers to process and store their catch aboard ship. Because they could remain at sea for longer periods of time, hunters took increased numbers of whales.

The Effort to Protect Whale Species

Few attempts were made to regulate whaling until 1930, when Norway formed the Bureau of International Whaling Statistics in order to investigate stocks and populations. In 1931, the International Convention for the Regulation of Whaling was established, making it illegal to hunt right whales or females with calves. Because this convention proved ineffective, a new regulatory system, the International Whaling Convention, was established in 1937. It prohibited the hunting of gray whales, defined a whaling season for the southern hemisphere, and set catch size limits for certain species, including the humpback. In 1938, specific humpback stocks were protected from pelagic whaling.

In 1946, the International Whaling Commission (IWC) replaced the International Whaling Convention. The IWC has since been the basis for the international regulation of all whale species. Charter members of the IWC included Argentina, Australia, Brazil, Canada, Denmark, France, Iceland, Mexico, The Netherlands, New Zealand, Norway, Panama, South Africa, the Soviet Union, the United Kingdom, and the United States. Japan joined in 1951, and Chile, the Republic of Korea, Peru, Seychelles, Spain, and Sweden became members in 1979. Not all whaling nations are members of the IWC, but about 90 percent of the world's catch of whales is taken by member nations.

The original objective of the IWC was to regulate whale stocks in the best interest of the whaling industry, which realized that its continued profitability depended upon healthy stocks. However, as time went on, member nations realized that populations of most species had declined so dramatically that extinction would occur unless the industry was put to an end. For example, a three year IWC study initiated in 1960 reported on populations in the Antarctic. Both the blue and humpback populations had seriously declined. The committee responsible for the study recommended full protection for these species, which was granted in 1966.

Public opposition in the United States led to the cessation of this country's whaling operations in the early 1970s. Riding the wave of popular sentiment, Congress passed several pieces of legislation concerning whales and other endangered animals. In 1971, the Pelly Amendment was added to the Fisherman's Protective Act. This amendment was a significant development, since it provides for an embargo on fisheries imports from any country conducting fishing operations that diminish the effectiveness of any international fishery conservation program.

In 1973, Congress passed the Marine Mammals Protection Act (MMPA), which protected 100 species within U.S. boundaries and initiated a moratorium on all marine mammals and trade in products without a permit. Penalties were instituted for all violations. The MMPA also established the National Oceanic and Atmospheric Administration (NOAA) and the National Marine Fisheries Service (NMFS). The Endangered Species Act, also passed in 1973, grants protection to all endangered creatures and their habitats.

The United States was not alone in its enthusiasm for conservation during the environmental awareness movement of the 1970s. Many other countries and organizations began to advocate the protection of endangered species. In 1975, this international concern culminated in the Convention on International Trade in Endangered Species of Wild Fauna and Flora (CITES). The result of nearly ten years of effort by the International Union for the Conservation of Nature and Natural Resources, CITES established rules for wildlife trade. Unfortunately, a compliance loophole allows a country to enter a "reservation" on a species, thus notifying other countries that it does not intend to comply with the trade restrictions on that species. For example, Japan submitted a reservation on the fin whale.

In 1982, the IWC approved, by a vote of 25-7, a moratorium on commercial whaling to take effect in 1985. Japan, Norway, Iceland, Peru, Brazil, the Soviet Union, and South Korea cast dissenting votes and filed formal objections. (Peru withdrew its objection the following year.) The success of the moratorium was compromised because the nations that harvest the majority of the whales did not agreed to comply. Moreover, as with CITES, a country can enter a reservation on a species.

Although Japan at first refused to agree to the moratorium on whaling, the United States applied economic pressure by invoking the Packwood-Magnuson Amendment to the Fisheries Protective Act in April of 1983. Passed in 1979, the Amendment stipulates that a nation that undermines any international effort to regulate whaling is subject to lose at least 50 percent of its fishing allocations in the U.S. fishery conservation zone. Japan lost 102,00 metric tons in fishery allocations in 1983 and another 20,000 metric tons in 1984. In 1984, Japan agreed to withdraw its objection to the moratorium if the United States would allow them to take whales until the end of the 1986/1987 whaling season without further sanctions. The United States agreed, and Japan officially ceased commercial whaling in 1987.

Nonetheless, many environmental groups and other member nations of the IWC are not satisfied with Japan's actions. They claim that Japan's Institute of Cetacean Research, which was formed in 1987, merely continues the whaling industry under the guise of scientific research. For example, the 273 minke whales taken as part of a 1988 "feasibility study" produced 1100 tons of whalemeat. Although Japan claims to market the meat as a byproduct, critics say it is the principal reason for the kill. Researchers at Cambridge University suggest that whales do not have to be killed in order for scientists to study such topics as social structure, paternity, and breeding patterns. Instead such information can be obtained through molecular analyses of small skin samples. Japan also purchases whalemeat from Iceland, the other country that continues to hunt whales in the name of research. Although the Japanese scientists say they must kill the whales in order to gather data about birth and death rates which will allow for better management of whale stocks, IWC scientists claim nonlethal methods could provide the same information.

LOOKING AHEAD: THE FUTURE OF THE BLUE, THE GRAY, AND THE HUMPBACK

The future of all whale populations to a large extent depends on how well they are able to recover from centuries of exploitation. Even with the protection most species have gained in recent years, it may be decades before we know whether enough animals of different species and stocks remain for recovery. Right whales have been protected in some way since the 1930s, yet until the 1970s there was virtually no evidence that their stock had rebuilt at all.

A 1989 census of whale populations in the Antarctic added to this uncertainty. The scientific committee of the IWC, which conducted the census, based their numbers on direct sightings, a method considered to produce more reliable data than previous estimates, which had been extrapolated from whale catches. Unfortunately, the new census revealed that Antarctic populations of many species were smaller than previously thought. For example, the blue whale population numbered 500, rather than 8000 (estimated in the 1970s). One exception was the humpback; its population proved to be 4000 rather than 3000, as previously thought.

The IWC moratorium, which is designed to continue indefinitely unless three-quarters of the member nations vote to overturn it, must continue if whale populations are to have any chance of full recovery. The "loophole" allowing countries to kill significant numbers of whales in the name of scientific research is a serious flaw in the moratorium, just as the objection system is a major weakness of the IWC. Because the commission is an international agency that lacks the authority to enforce its recommendations, nations can ignore any regulations with which they disagree. Because economic incentives are typically at the root of modern whaling operations, economic disincentives are the most effective tool we can use to end whaling. Without economic pressure applied by nonwhaling countries, whaling nations will continue to ignore the IWC moratorium. In addition, individual citizens can apply economic pressure on whaling nations through actions such as the Greenpeace boycott of Icelandic fisheries products.

The New Management Procedure, instituted by the IWC in 1974, provides effective and sound management of all whale stocks. The IWC should reinstitute this plan at the end of the current ten-year moratorium. Further research, especially the estimation of populations, should improve the procedure. But even the best management plan cannot safeguard a resource that has been depleted beyond its ability to recover. Unless the moratorium is successful and whale stocks begin to rebound, the question of their management may be moot by the end of this century.

Before we can safeguard whale stocks from indirect pressures that degrade their habitat, intensive research must be conducted to determine the effects of various human activities. Because we know relatively little about the whales, especially about their migratory routes, it is difficult to determine which areas need protection, the size of the areas required to safeguard the routes, and the exact times of the year routes should be safeguarded.

SUGGESTED READING

"A Way of Counting without Killing." *New Scientist*. July 2 1987. 30-31.

Allen, Kenneth Radway. *Conservation and Management of Whales*. Seattle: Washington Sea Grant (distributed by University of Washington Press), 1980.

Anderson, Alun. "More Insults Fly but Japan's Scientific Whaling Goes On." *Nature*. February 9 1989. 494.

Andrews, Roy Chapman. *All About Whales*. New York: Random House, 1954.

"As the Chain of Death Spreads to Whales." *New Scientist*. January 28 1988. 30.

Burton, Robert. *The Life and Death of Whales*. New York: Universe Books, 1973.

Cherfas, Jeremy. "Time to Come Clean on the Whales." *New Scientist*. March 25 1989. 63-64.

Cousteau, Jaques-Yves, and Yves Paccalet. *Planete des Baleines*. Translated by I. Mark Paris. Hill Street, London: W.H. Allen, 1988.

Day, David. *The Whale War*. San Francisco: Sierra Club Books, 1987.

Dietz, Tim. *Whales and Man: Adventures with the Giants of the Deep*. Dublin, NH: Yankee Books, 1987.

Ellis, Richard. "A Sea Change for Leviathan." *Audubon*. November 1985. 62-79.

———. *The Book of Whales*. New York: Knopf, 1980.

Ellis, Sara L. "Japanese Whaling in the Arctic: Science of Subterfuge?" *Oceanus*. Summer 1988. 68-69.

Evans, Peter G.H. *The Natural History of Whales and Dolphins*. London: Christopher Helm, 1987.

Gambell, Ray. *Whales*. New York: Mallard Press, 1989.

Hall, Howard. "Focusing on a Blue . . . Giant." *International Wildlife*. July/August 1989. 30-33.

Hall, Sam. "Whaling: The Slaughter Continues." *The Ecologist*. 1988. 207-12.

Horgan, John. "A Grave Tale: Do Whale Remains Help Life Spread on the Deep Sea Floor?" *Scientific American*. January 1990. 18+.

Hoyt, Erich. *Seasons of the Whales: Riding the Currents of the North Atlantic.* Post Mills, VT: Chelsea Green, 1990.

Hunter, Robert. *To Save a Whale: The Voyages of Greenpeace.* San Francisco: Chronicle Books, 1978.

Hyman, Randall. "Old Enemies in the Same Boat." *International Wildlife.* July/August 1990. 12-16.

Jones, Mary Lou, Steven L. Schwartz, and Stephen Leatherwood, eds. *The Gray Whale: Eschrichtius robustus.* Orlando, FL: Academic Press, 1984.

Lowenstein, Jerold. "Cetacean Evolution." *Oceans.* January/February 1986. 70-1.

Martin, Anthony. *Whales.* New York: Crescent Books, 1990.

McIntyre, Joan. *Mind in the Waters: A Book to Celebrate the Consciousness of Whales and Dolphins.* New York: Scribner, 1974.

McNulty, Faith. *The Great Whales.* Garden City, NY: Doubleday, 1974.

Miller, Tom. *The World of the California Gray Whale.* Santa Ana, CA: Baja Trail Publications, 1975.

Mowat, Farley. *A Whale for the Killing.* Boston: Little, Brown, 1972.

Payne, Katy B. "A Change of Tune (Songs of the Humpback Whales)." *Natural History.* March 1991. 44-46.

Small, George L. *The Blue Whale.* New York: Columbia University Press, 1971.

Stone, Gregory. "Cetacean Prognosis." *Oceans.* November/December 1988. 56-57.

Taylor, Michael. "Sound Strategies for Survival." *New Scientist.* October 30 1986. 40-43.

"Voice of the Whale." *Oceans.* September/October 1987. 5.

Walsh, John. "Research Whaling on the Table." *Science.* July 31 1987. 481.

Watson, Lyall. *Whales of the World.* London: Hutchinson, 1985.

Watson, Paul. *Sea Shepherd: My Fight for Whales and Seals.* New York: Norton, 1982.

Wexo, John Bonnett. *Whales.* San Diego: Wildlife Education, Ltd., 1983.

"Whale Populations Take a Dive." *Environment.* July/August 1989. 24.

Winn, Lois King. *Wings in the Sea: the Humpback Whale.* Hanover, NH: The University Press of New England, 1985.

CHAPTER 5

APPLYING ECOLOGICAL PRINCIPLES: PRAIRIE RESTORATION AT FERMI NATIONAL ACCELERATOR LABORATORY

Like the evening news, discussions of environmental science tend to dwell on the negative: depleted resources, endangered species, predicted catastrophes. Environmental science explores the interaction between human culture and the natural world, and unfortunately many of these interactions somehow damage the delicate balance of natural systems. However, there is one aspect of environmental science that, although necessitated by the destructive behavior of the human population, does generate more positive news. Applied ecology is the scientific discipline that attempts to predict the ecological consequences of human activities and to recommend ways to limit damage to, or even to restore, natural ecosystems. Applied ecologists work to understand the stresses to which ecosystems are exposed as well as the ways in which natural systems respond to, recover from, and adapt to disturbances. In addition, the research of applied ecologists can give us a basis on which to plan solutions to environmental problems.

We must keep in mind, however, that any plan for environmentally sound management should integrate a knowledge of natural systems with an awareness of human institutions and social considerations. Cultural factors such as politics, economics, ethics, and environmental education have a profound effect on how and if environmental problems are addressed. Thus, the exciting new approaches to managing ecosystems developed in applied ecology can achieve their full potential only in a society that is environmentally aware and responsible.

Although applied ecologists often use their knowledge of natural systems to conserve or protect habitats or specific species, in some cases they work to actually restore or reconstruct ecosystems. The goals of restoration ecology, as this field is called, are:

1. to repair biotic communities whenever possible, or to reestablish them on their original site (if the community is destroyed), or to replace those communities on other sites if the original sites can no longer be used;
2. to maintain the present diversity of species and ecosystems by finding ways to preserve the biotic communities or to protect them from human disturbances so that they can evolve naturally; and
3. to increase our knowledge of biotic communities.

A specific restoration project often yields knowledge that can then be applied to help restore or manage other systems. A variety of ecosystems have been the focus of restoration projects, including holdings in state and national forests, phosphate mines in Florida, coal strip mines in Pennsylvania and Illinois, and freshwater and marine reefs.

Another endangered ecosystem that holds special significance for the United States is the prairie. In American history and imagination, the prairie is often pictured as an "inland sea" of waving grass stretching to the horizon, grazed by herds of buffalo and traversed by wagon trains heading west. In reality, however, 99 percent of the great prairie that once covered most of the Midwest is gone, paved over by cities and supplanted by farms that have flourished on its rich, deep soil. Most of us have never seen a prairie and would have difficulty distinguishing a patch of this unique ecosystem from an overgrown agricultural field.

In addition to playing an integral part in American culture, the midwestern prairie is the only habitat where a vast array of grasses, flowers, insects, birds and mammals can be found. Despite its unique value, the prairie once seemed doomed to survive only in small patches, and even those were threatened by developers and nonnative weeds. Fortunately, a growing number of ecologists have begun to restore examples of the prairie community, and their efforts hold out the hope that this magnificent and self-renewing landscape may once more become a familiar sight. The largest and one of the oldest prairie restoration projects is taking place on the grounds of Fermi National Accelerator Laboratory (Fermilab), a nuclear physics research institution near Chicago, Illinois. This chapter explores the ways this project illustrates both the difficulties and the promise of restoration ecology.

PHYSICAL BOUNDARIES: THE DISAPPEARANCE OF THE TALLGRASS PRAIRIE

The original North American prairie, which once stretched from Indiana to Nebraska, was composed of three types of grass: tallgrass, mixed-grass, and shortgrass. Farthest to the east, where heavier rainfall allowed more lush growth, the tallgrass prairie covered a sort of triangle with its points in northeastern Indiana, Saskatchewan (Canada), and Texas. The prairie landscape included more than grassland—prairie marshes and oak savannas (wooded areas with clear, grassy understories) were prominent communities as well.

Illinois, still known as the Prairie State, occupied the heart of the tallgrass region. Forty thousand square miles of prairie landscape covered a full 70 percent of the state. Today, only four or five square miles of prairie are left intact. Most of this exists in small patches, primarily along railroad tracks or in pioneer cemeteries. The larger patches, which were usually left unplowed and undeveloped because of poorer soils, are not representative of the deep, rich loam that characterized the original prairie community.

In 1972, the Department of Energy (DOE) commissioned the construction of Fermi National Accelerator Laboratory on almost 7000 acres of farmland (originally prairie) 30 miles west of Chicago. Managed by a consortium of universities for the DOE, the lab concentrates on investigating the nature of protons. Their huge proton accelerator, a ring four miles in circumference buried beneath the earth, contains one thousand powerful magnets that force protons around the ring at 99.99 percent of the speed of light. When the protons hit their targets, they break down into smaller components.

In the scientific tradition, Fermilab tries to break the proton down into its smallest parts in order to fathom its workings. A contrasting experiment is occurring on the 640 acres within the accelerator ring: dedicated ecologists, including Robert Betz of Northeastern Illinois University, Fermilab staff, and volunteers, are working to build a prairie ecosystem out of its constituent parts—soil and prairie plant species—in order to preserve the community and all its complex interrelationships. The Fermilab site also boasts archaeological significance due to the discovery of 25 sites showing evidence of more than one prehistoric culture.

BIOLOGICAL BOUNDARIES: THE BIOTA OF THE TALLGRASS PRAIRIE

Although the Fermilab Prairie is much smaller and less diverse than the climax Illinois prairie found by the settlers, a knowledge of the complex relationships among the animals and plants of the original prairie ecosystem is invaluable for those hoping to restore that ecosystem.

The Climax Illinois Prairie

Although virtually every large stretch of original Illinois prairie was destroyed before botanists could study it, the team at Fermilab and other prairie experts in the area pieced together as much information as they could from local prairie remnants. Perhaps the most valuable of these, the 100-acre Gensburg-Markham Preserve, was discovered just thirty years ago in Markham, a suburban area south of Chicago. This high-quality prairie was slated for development in the early 1900s, but escaped because of the economic collapse of the Great Depression. This reprieve would have been temporary were it not for the efforts of Betz, residents of the Markham area, and the Nature Conservancy. As as a result of these efforts, the land was either donated to or bought by the Nature Conservancy, which in turn passed it along to Northeastern Illinois University.

Gensburg-Markham has never been plowed (with the exception of a few acres), and scientists believe it has changed little since the last glacier receded 8000 years ago. In 1988, the National Park Service designated Gensburg-Markham a National Natural Landmark. Besides its inherent value as a rare habitat for endangered species that can live only on high-quality prairies, Gensburg-Markham has proven to be an invaluable source for information and prairie plant seeds for the Fermilab project.

The prairie community exemplified by Gensburg-Markham is exceptionally self-renewing and reliant; while maintaining its impressive diversity of plants and animals, it survives all extremes of weather, from parching droughts to sub-zero winters. One reason for this extraordinary endurance is the extensive root networks of the plants. Although it may grow only 3-6 inches above the surface in its first year, a prairie grass like the big bluestem can simultaneously develop a root system three times that size. A full-grown big bluestem's roots might be two feet across and five feet deep. Many plants also reproduce by putting out rhizomes (horizontal root-like stems), from which new plants sprout. The rhizomes and roots all grow together into a dense tangle that holds the soil in place, helping it to retain moisture and preventing erosion. This tangle also slows decayed matter from washing through the soil, thus encouraging new soil to accumulate. By the time settlers arrived at the prairie in the 1830s, the soil was thicker and richer than any they had ever encountered.

Some prairie grasses and flowers emit substances that prevent or enhance the success of other plants. In this way, the prairie plants help to form a soil conducive to the more rare and delicate prairie species while discouraging the growth of "weedy" species not native to the prairie.

One important factor needed to maintain a prairie community cannot be provided by the prairie itself. Although full of lush growth during the summer, the prairie very quickly dies back in the fall. These vast expanses of dried grass provided ideal tinder for fires started by lightning, the spontaneous combustion of organic matter, and Native Americans, who learned to light them to flush out animals or improve grazing for the elk and bison that were their prey. Before settlers arrived and began to fight these prairie fires, huge walls of flame used to rage across the grasslands regularly, sometimes lasting for days and burning hundreds of square miles.

Far from being damaged by such blazes, prairie plants were fire-dependent—in other words, they had evolved over millions of years not only to survive but actually to require frequent fires. With their life-supporting roots safe underground, they could easily regenerate in the spring. In fact, by burning away the dead growth of the previous year, prairie fires allowed more sunlight to reach the new growth when it sprouted. Fire also helped to recirculate nutrients into the soil by turning dry and dead vegetation to ash, and even helped to propagate certain plant species that needed the fire's stimulus to release seeds or produce new growth. Most prairie animals survived the fire by burrowing beneath the ground or running away. Nonprairie plants, especially trees that would begin to invade the grassland regions, were exterminated by the fire with the exception of the fire-resistant oak trees of the savanna regions. Thus the seemingly destructive fires actually played an integral role in preserving the prairie's unique nature.

And the prairie's biotic community is indeed unique. Researchers estimate that the original Illinois prairie supported roughly 200 species of plants and 800 species of insects, birds, and mammals, all dependent on each other. Many prairie species can survive nowhere else; they require the blend of chemical, physical, and biological interactions found only in the prairie. The Gensburg-Markham Preserve is home to many of these rare species—including the prairie orchid, the Aphrodite butterfly, and the smooth green snake—although its small size prevents it from housing the full array of prairie inhabitants.

Prairie plants can usually be classified as grasses, forbs (wildflowers), or legumes. The colorful forbs are the most eye-catching. Imagine an assortment of goldenrod, purple gentians, blue asters, white false indigo, yellow black-eyed Susans, and pink blazing stars. Although fewer in species and less spectacular, grasses dominate the prairie in numbers. The taller grasses such as big bluestem and Indian grass grow to between six and twelve feet tall. Many plants are uniquely adapted for prairie conditions. The compass plant, for example, always orients its leaves east-west in order to expose them as little as possible to the sun and thus to save water. In the savanna areas of the prairie, tree species are primarily oak, hickory, and white ash. Marshy areas have their unique species, too, such as prairie sedge and water parsnip.

These plants support a huge insect population, many of which are dependent on one or two particular species of plant for survival. For example, the Baptisia dusky wing butterfly feeds only on wild and false indigos. Species of worms specific to the prairie help maintain its unique soil characteristics. Approximately 300 species of birds are at home on the prairie, although many are migratory and spend just part of the year there. Prairie birds include bobolinks, eastern meadowlarks, savannah sparrows, grasshopper sparrows, Henslow's sparrows, and upland sandpipers. Prairie mammals include meadow voles, groundhogs, mink, beaver, muskrats, rabbits, coyote, and deer. Although many of these species have successfully adapted to man's alteration of the prairie and can live elsewhere, other prairie animals such as bison, wolves, elk, and bobcats have completely disappeared from the Illinois area.

The Fermilab Prairie

The prairie at Fermilab is a young prairie, just as a baby is a young human or a sapling is a young tree; it still has a long way to go before it resembles the climax prairie community. However, it does have the advantage of large size, size enough to become a real, functioning ecosystem. Most of the original prairie remnants are just too small to support the full range of species and activities that occurred in the presettlement Illinois prairie.

As Betz and his colleagues attempted to create a truly prairie-sized Illinois prairie, they discovered, largely by trial and error, that a new prairie must be established in stages. Although they first planted approximately seventy prairie species on the Fermilab site, only about 25 plants—what they call the "prairie matrix"—really flourished. Not surprisingly, these same plants are the last to disappear when a piece of prairie is invaded by "weedy" species. Many of these weeds are Eurasian in origin, brought to the Midwest by settlers. Without the benefit of fire, prairie plants gradually lose their hold to species like Hungarian brome, ragweed, and Queen Ann's Lace.

The first year that prairie plants grow in a plot, they are barely evident, almost completely overshadowed by the nonprairie "weeds." After three or four years, the matrix plants have established deep, complex root systems and begin to compete with the weedy species in size and number. Indian grass and big bluestem are two highly competitive prairie grasses that do very well in those first years. By the fourth year, the prairie plants are ready to sustain a fire, which the Fermilab team usually sets in the early spring, after some of the weedy species have begun to germinate but before the prairie plants begin to grow. After the first fire, the prairie matrix definitely gains the upper hand and begins to push out its competitors.

Eventually, the matrix plants alter the soil composition and chemistry enough to enable a second wave of less hardy prairie plants to grow, including sky-blue, smooth, and heath asters, cream wild indigo, blazing stars, mountain mint, and prairie dropseed. Sometimes, seeds of these second-wave plants—sowed ten or fifteen years before—suddenly sprout, indicating that something important has changed in the soil. Betz and his colleagues, however, are occupied with the practical business of nurturing a prairie, and leave it to future research to discover exactly what has changed and why. After the first- and second-wave plants have modified the soil to almost presettlement conditions (a process that will probable take between 50 and 100 years), the prairie will be ready for the most delicate, rare species such as milkwort gentians.

SOCIAL BOUNDARIES: CHANGING ATTITUDES TOWARD THE PRAIRIE

The speed with which the 40,000 square mile Illinois prairie shrank to remnants totaling 4 or 5 square miles is a testament to the early settlers' drive to move westward and to make productive use of their new found land. The current drive to reconstruct prairies and other ecosystems where they once flourished represents an equally adventurous spirit, but this one is a spirit of preservation and stewardship.

When settlers first encountered the tallgrass prairie, they quickly learned that its deep, rich soil was ideal for cropland and that the flat terrain with few rocks made plowing even easier. With the introduction of drainage tiles around 1900, even those areas that had been too marshy to farm were transformed into arable land. Quietly, the seemingly endless prairie gradually disappeared beneath the plow and the construction of cities like Chicago.

Restoration ecologists like Betz who attempt to restore prairies face a monumental task—the idea of restoring ecosystems is relatively new, and they must make their way by trial and error. The process is laborious, time-consuming, and so slow that those who begin a prairie will probably not live to see it in its glory. Nevertheless, more and more people are devoting themselves to reconstructing this fascinating landscape that once covered our midwestern plains for a number of compelling reasons.

The first of these reasons is that the prairie is our national heritage. The bounty of its rich soil has allowed our farms to flourish, and the animals which inhabited it helped to feed and clothe the early settlers. The boundless sea of grass stretching to the horizon, although it no longer exists, remains an integral part of our nation's history, symbolizing the freedom and openness which we prize so highly.

The prairie ecosystem is also valuable for scientific reasons. Many of the plants and animals that live only on the prairie would become extinct if their habitat disappeared. These threatened species hold an untold wealth of genetic knowledge that we might use to create better crops or produce new medicines. Scientists still have much to learn about how the complex prairie ecosystem developed and how it interacted with its physical environment. For example, the way that the prairie's perennial polyculture discourages weeds, damaging insects, and diseases while accumulating rather than losing soil could provide us with clues for developing sustainable agriculture.

Finally, many of those involved in prairie restoration firmly believe they have a moral obligation to preserve the prairie ecosystem and the species that compose it. Each tiny insect, rare flower, and waving grass is an important part of the environment and has an inherent right to survive. In the case of many prairie species, every chance for survival will disappear unless prairies large enough to function as ecosystems are restored or preserved.

LOOKING BACK: THE HISTORY OF THE FERMILAB PRAIRIE RESTORATION PROJECT

The restoration ecology movement has its roots in the growing realization that we cannot go on using and altering our environment without protecting and even restoring certain areas. This realization began to take hold after the Great Depression, when catastrophes like the Dust Bowl dramatically illustrated the consequences of abusing our natural resources. The tallgrass prairie has been the focus of many of the first restoration projects, primarily because so little of it is left.

When Betz and the Fermilab Prairie Committee first began their prairie restoration project, they followed a path blazed by only a handful of pioneering ecologists. One of the first major voices in the ecology movement, Aldo Leopold, helped to catalyze the first major prairie renovation at the University of Wisconsin Arboretum in 1936. Dismayed at the complete alteration of the landscape, he wanted to "reconstruct a sample of original Wisconsin— a sample of what Dane County looked like when our ancestors arrived during the 1840s." The result, the Curtis Prairie, is a 60 acre tract largely created by transplanting sod from prairie remnants. Using more sophisticated techniques, Henry Green began the 35-acre Green Prairie at the Arboretum in the early 1940s. Today, parts of the Curtis and Green Prairies are fairly similar to the original prairie ecosystem. Another major figure in prairie restoration was Ray Schulenberg of the Morton Arboretum near Chicago. In 1962, he began restoring a prairie there, growing prairie species in flats and planting them by hand.

Despite such successful projects, the theory behind prairie restoration is still largely unexplored. Even the knowledge of the ecological succession of the prairie and how to build such an ecosystem is incomplete, and pioneers like Schulenberg and Betz have had to proceed by trial and error. The nature of prairie restoration requires a choice between putting effort into actually growing a good-sized prairie or experimenting to find out why the prairie grows the way it does. So far, ecologists seem to have made the first choice, leaving the second area for more traditional research scientists. Although few such research projects have been undertaken so far, the growing interest in ecosystem restoration may spur those needed studies in the future.

For Betz and the others involved at Fermilab, the top priority was to create the entire ecosystem, not simply to copy of the work of their predecessors. Betz envisioned a restored prairie on a previously unheard of scale: hundreds, perhaps thousands of acres that could support the full range of prairie species and interact as a true ecosystem. Although the traditional scientific method would insist that extensive testing be conducted before any large tracts be planted, restoration ecologists question whether experiments in small plots can predict behavior on an ecosystem level. Besides, the creators of the Fermilab prairie believed that in order to demand the attention it deserved, it should be as large as possible right away.

Visions of the Future

The unusual combination of nuclear physics and prairie restoration at Fermilab grew out of the visions of two men: Robert Betz, professor of biology at Northeastern Illinois University, and Robert Wilson, first director of Fermilab. Betz became interested in the native Illinois prairie during a 1959 field trip to a prairie remnant with Floyd Swink, author of *Flora of the Chicago Region*. He built on that interest by helping Ray Schulenberg to grow his prairie seedlings at the Morton Arboretum. Betz also gained permission to manage prairie remnants on some cemeteries that dated back to settlement days, using them as ecological laboratories to which he added missing prairie species and removed nonprairie plants.

In the early seventies, after approximately a decade of working with prairie species, Betz heard that Robert Wilson, director of the newly created Fermilab, had approached the Morton Arboretum for ways to landscape the thousands of acres owned by the lab. Seeing those thousands of acres as a unique opportunity to restore an entire ecosystem, Betz and Schulenberg set up a meeting with Wilson.

Betz describes Wilson as a man of vision who was already interested in the idea of prairies. Wilson had even purchased a herd of bison to raise on the lab grounds. Their meeting was "like a positron and an electron hitting together," and the result was a huge burst of energy that would fuel the largest prairie restoration project in the country.

Once the restoration project received official backing from the Nature Conservancy, the DOE approved the plan. Fermilab set up a Prairie Committee made up of Fermilab staff to manage the project, and allocated equipment and the time of its roads and grounds staff for prairie work. Betz and Schulenberg originally served as advisors to the committee, and Betz has remained the guiding force behind the project.

Building the Prairie

Since no one had ever attempted a restoration on the scale of the Fermilab project, Betz and his team had to develop new ways of planting and managing the prairie. Hand harvesting and planting were simply impractical for a prairie that would eventually cover hundreds, perhaps thousands, of acres.

The first step occurred in the fall of 1974. By hand-collecting prairie seed from local remnants, almost 100 volunteers gathered 400 pounds of seeds from over 70 species. The next spring, those seeds were planted on 9.6 plowed and disked acres within the accelerator ring. Although weeds dominated in the first year, the hardier prairie plants eventually gained a foothold. The first burning occurred in 1978. By the mid-1980s, virtually all the area available for cultivation within the ring (385 acres) was planted with prairie species. The ring also boasts an open bur oak forest and some marshy areas, just like an original prairie landscape. Eventually, the planting extended outside the ring, bringing the total acreage to 640 as of 1990. This includes a 90 acre plot with an interpretive trail named for Margaret Pearson, Fermilab's public information officer and a fervent supporter of the prairie project. The trail is open to the public and includes signs explaining the prairie ecosystem and the prairie restoration process.

Over the years that they have been working on the prairie, Betz and his colleagues have gradually evolved successful techniques for harvesting and planting which use farm equipment such as plows and combines—ironically, the same equipment that helped to destroy the original prairie. Many of the techniques used on the Fermilab prairie were experimental when first tried, such as the idea of using the combine to harvest prairie seed. Over the years, however, the Fermilab project has developed successful, efficient methods for establishing prairie plants over large areas, and these methods can now serve as models for other restoration projects.

The entire planting process begins in the fall, when the area to be seeded is mowed and then plowed. Exposing weed roots to the freezing temperatures helps to discourage their growth the next year. The following May, the area is disked and harrowed to level and dislodge roots, and then "cultipacked" to tamp down the soil slightly, providing a firm surface for the seeds. Once the seeds are spread, the cultipacker runs over the soil to press them into it slightly. Given two weeks of good weather, the crew can prepare and plant approximately 70 acres with prairie seed.

For planting, the crew first tried a Nisbet drill to sow the seeds, but this method left obvious rows and was slow and inefficient for large areas. After four sowings with the drill, it was replaced with a hydromulcher, which sprayed the seeds out in streams of water but was so weighty it bogged down in the soil. A salt-spreader truck did a fairly good job of scattering the seeds quickly, and an all-terrain spreader (normally used to spread fertilizer), which can hold 1000 pounds of seed, was an even greater improvement due to its wider spreading arc and greater maneuverability.

The crew uses a combine to harvest 5,000-10,000 pounds of seed from older plots on the Fermilab prairie each autumn. Volunteers still hand-collect 100-200 pounds from remnant prairies to increase genetic variability, adding new species and supplementing those already at Fermilab in small numbers.

The Fermilab project has also been one of the innovators in the use of fire to aid in prairie restoration. During the 1930s and 1940s, the use of fire in any ecosystem was widely discouraged by ecologists. When the staff of the University of Wisconsin Arboretum began to burn their prairies in the early 1950s, they faced a great deal of criticism. However, fire has proved invaluable to restoring and maintaining prairies; studies show that burned prairies produce twice as much living matter as those that have been left unburned for even a few years. Without burning, constant weeding is required to keep weedy species and trees from invading the prairie. Betz and Schulenberg were largely convinced to try fire based on historical accounts of prairie fires in the area. Though most natural fires occurred in autumn, they found that burning in the spring, after the weedy species have begun to germinate but before the prairie plants get started, works best.

While still overpopulated with grasses like big bluestem and infested with non-prairie species like white sweet clover, the Fermilab site boasts approximately 275 species of prairie and marsh plants. In the older tracts, second-wave species are beginning to thrive, although many exist only in small numbers, and in some areas along service drives, big bluestem is even beginning to invade without the help of lab staff.

Of course, establishing the plant species is just the beginning of restoration. A prairie is also made up of the many species of wildlife that live there. Although Fermilab still lacks many prairie species, especially invertebrates, other animals are making an impressive comeback. Coyotes, deer, minks, and weasels can be found on the site, and the frog population also seems to be growing. As of 1989, a total of 224 species of birds had been sited, including two from the federal list of endangered species (the peregrine falcon and the piping plover) and fifteen from the state list. The endangered sandhill crane is actually breeding on the site, and the long- and short-eared owls, also endangered in Illinois, spend time there. Some rare trumpeter swans have been brought in to live on the pond within the accelerator ring.

Research on the Fermilab Prairie

As interest in restoration ecology grows, projects like the Fermilab Prairie attract more and more scientists interested in researching the complex processes and interactions that underlie the laws of prairie plant succession. For example, a group of scientists from Argonne National Laboratory conducted a study designed to investigate the redevelopment of the soil by comparing it to soil in native prairies, fields, and cultivated fields. They found that the soil on the Fermilab site showed a marked improvement after 10 years of prairie growth: as the prairie plants grew, the soil developed into aggregates, or clumps held together by organic material. The scientists hypothesized that these aggregates helped make the soil more resistant to erosion and better able to accumulate.

Other projects include a survey of the birds associated with the Fermilab prairie. Of the more than 200 species spotted, over 40 actually bred on the prairie. Other surveys concentrated on populations of insects, reptiles, amphibians, and small mammals. Such projects are invaluable for several reasons. First, they let the Fermilab team know what species still need to be restored. Second, such studies are an impressive measure of the project's success to date because they are indications that the goal of the project—to establish a healthy, holistic prairie ecosystem—is being achieved.

LOOKING AHEAD: THE FUTURE OF THE FERMILAB PRAIRIE AND PRAIRIE RESTORATION

The Fermilab prairie restoration project has flourished under the dedicated direction of Betz and with the generous support of the lab's directors and staff. The Fermilab project promises to become a close representative of the original Illinois prairie, and to provide other restoration projects with proven methods of establishing a prairie community on a large scale. It also holds the potential for endless research projects, from simple inventories of animal and plants species to investigations into the complex biochemical processes that regulate the ecosystem.

Continuing the Restoration

In the next ten years, Betz and the Fermilab Prairie Committee plan to work toward a number of important goals. First, they will continue to add new plots of prairie plants every year. Planting plots in chronological order provides a wonderful opportunity for researchers who wish to compare various stages of succession, while annually adding to the prairie's total area. At the same time, the committee plans to maintain their effort to reintroduce second-wave and the even less tolerant third-wave species to the older plots. They also plan to enrich marsh areas with original dominant species such as great bulrush and swamp dock. Their goal is to eventually link all the separate plots outside the ring into one large, contiguous prairie landscape, including marsh and savanna, so that the original complex interaction of ecosystems can occur.

Another major stage in the restoration will be the reintroduction of animal species. Although many other bird species have successfully reintroduced themselves, the committee is interested in establishing the prairie chicken at some point. Franklin's ground squirrel, successfully reintroduced to Gensburg-Markham, is another important species they hope to acquire. Insects, however, present perhaps the greatest challenge. Because most invertebrate species cannot travel far enough from prairie remnants to reach the Fermilab prairie, a comprehensive program will have to be undertaken to establish populations of prairie earthworms, yellow-winged grasshoppers, and countless others.

Potential for Research

In 1987, the U.S. Department of Energy designated Fermilab as a National Environmental Research Park, one of a network of DOE sites used to study, predict, and minimize the effects of human stresses on unique ecological communities. Each of the six parks currently in the network contains ecosystems unique to a geographical region of the United States. Other parks include the Hanford Reservation in Washington, the Savannah River Plant in North Carolina, and Los Alamos National Laboratory in New Mexico. Fermilab, of course, represents the threatened midwestern prairie grassland and oak-savanna ecosystems. By far the most open and accessible site in the system, Fermilab offers a special opportunity for public education. Designation as a NERP further confirms Fermilab's value as a research site, and should encourage scientists to select the site for diverse studies.

Proposed research projects include a comprehensive characterization of the developing prairie and other communities on the site and long-term studies to document successional changes in the developing communities. The Fermilab prairie would be especially interesting to those studying succession because it illustrates how human intervention can help a climax plant community evolve much more quickly than it would have naturally. Another promising research area is the effect of antibiosis (the process by which plants modify their environment through the production of various chemicals) on competing nonprairie plants. Projects that would be especially valuable to those restoring prairies might concern how to shorten the process of ecological succession and how to successfully re-introduce animal populations.

Growing Interest in Prairies

Although few people had heard of prairie restoration when Fermilab first offered its support to Betz in 1974, interest is rapidly growing. As more and more people become concerned about the environment and endangered species, the project attracts more visitors. The group of eager volunteers swells every year.

In addition to this growing public interest, prairie restoration is beginning to attract the attention of the government and private industry as a method of land management. Because the prairie can regulate itself with the help of a controlled fire every few years, it is an economical, easy way to manage large tracts of land while providing an interesting, beautiful example of a native ecosystem. State governments have begun to plant prairie species along roads and on highway medians, and major corporations like General Electric's Medical Care Systems Division in Milwaukee, Wisconsin, are beginning to grow prairies on their grounds. General Electric installed an 80-acre prairie for less than half of what it would have cost for a traditional bluegrass lawn. In addition, the prairie costs just $5 per acre per year to maintain.

Will the Fermilab prairie and the other restored prairies ever be identical to the original ecosystem? Opinions vary, although the Fermilab project has accomplished in 20 years what it might have taken nature 40 or 50 to achieve. Some scientists doubt if every weedy species will ever be eradicated, and an untold number of species may already be extinct. However, if the majority of prairie species and a few nonnative species can interact as a self-sustaining ecosystem, they will act as a reasonable substitute for the native prairie.

Although many restoration ecologists feel strongly about preserving endangered plants and animals for the inherent worth of those species, they also recreate these lost ecosystems for their fellow human beings. They hope that, by doing so, they will foster an appreciation for the complexity and beauty of nature. If they can do that, then perhaps future generations, unlike their ancestors, will coexist with the new prairies. Although restored prairies can never replace the original ones, they will have their own special symbolism. They will always represent one of the first ways our society tried not only to halt our degradation of the environment, but actually to repair it.

SUGGESTED READING

"Atoms and Animals Coexist at Fermilab." *International Wildlife*. November/December 1984. 27.

Berger, John J. *Environmental Restoration: Science and Strategies for Restoring the Earth*. Washington, DC: Earth Island Press, 1990.

——. "The Prairie Makers." *Sierra*. November/December 1985. 64-70.

——. *Restoring the Earth: How Americans Are Working to Renew Our Damaged Environment*. New York: Alfred A. Knopf, 1985.

Betz, Robert F. "One Decade of Research in Prairie Restoration at the Fermi National Accelerator Laboratory (Fermilab)." *Proceedings of the Ninth North American Prairie Conference*. July 29-August 1, 1984, Moorehead, Minnesota.

Brown, Kevin A. "Fermilab Joins a Network of DOE National Environmental Research Parks." *Fermilab Report*. March/April 1989. 18-25.

——. "The Fermilab Prairie: A Functioning Ecosystem." *The Fermilab Report*. May/June 1988. 33-39.

Buckley, G.P., ed. *Biological Habitat Reconstruction*. London: Belhaven Press, 1985.

Docekal, Eileen. "Tale of the Tallgrass." *Sierra*. May/June 1987. 76-9.

Fermilab National Environmental Research Park Program Plan, October 1987.

Fermilab Prairie Committee Ten-Year Plan. "Executive Summary." April 3 1989.

Jordan, William R., III., Michael E. Gilpin, and John D. Aber, eds. *Restoration Ecology: A Synthetic Approach to Ecological Research*. Cambridge: Cambridge University Press, 1987.

Kanfer, Larry. *Prairiescapes: Photographs*. Urbana, IL: University of Illinois Press.

Mlot, Christine. "Restoring the Prairie." *BioScience*. December 1990. 804.

Pemble, Richard H., Ronald L. Stuckey, and Lynn Edward Elfner. *Native Grassland Ecosystems East of the Rocky Mountains in North America: A Preliminary Bibliography*. Grand Forks, ND: University of North Dakota Press, 1975.

Sullivan, Jerry. "Bringing Back the Prairie." *Audubon*. July 1988. 40-47.

Sutton, Christine. "Home on the Nuclear Prairie." *New Scientist*. August 1 1985. 36-39.

Thomsen, Dietrick. "The Lone Prairie." *Science News*. October 16 1982.

CHAPTER 6

HUMAN POPULATION DYNAMICS: ETHIOPIA

If we see environmental problems such as dwindling resources, eroding soil, and growing pollution as the results of human actions, we must ask ourselves how such problems will be affected by the fact that the human population is rapidly growing. In 1989 the world population reached an estimated 5.2 billion. Growing at a rate of 1.8 percent, the population adds almost 90 million people annually—roughly equivalent to adding a country the size of Mexico each year. More people were added to the global population in 1990 than in any previous year in the history of humankind. Moreover, population growth is expected to increase each year of the 1990s, in effect setting a new record for growth with each passing year! If this rate of growth remains constant, world population could reach over 10 billion by about 2025.

Some people believe that this unchecked growth of the human population is actually a positive development. If people are the world's ultimate resource, they argue, then the more of them, the better. Proponents of this view believe that a growing population can make a country more prosperous because more people add producers and consumers to a country's economy. They believe that human inventiveness will solve any problems—environmental, economic, or social—associated with the increased numbers of people on the earth.

Many other people point out that, although the image of each child as a potential gift to humanity is a powerful one, the realities of population growth and starvation in countries such as Ethiopia and Bangladesh belie the idea that unchecked growth is unproblematic. Can those who have little food or water, no home, and the specter of a bleak future fulfill their human potential? Proponents of this view see continued population growth as a threat to environmental quality, political stability, international relations, and continued economic development. They contend that even if the rate of population growth could be decreased immediately, it would take many years to slow and halt the total increase in the absolute numbers of human beings on earth. In the meantime, the human population would continue to stress and damage the earth's life support systems, with dire consequences for the biosphere and its human and nonhuman inhabitants.

Although a variety of views on population growth exist, environmentalists agree that the human population's current use of environmental resources is unwise. We are consuming the "capital" of the biosphere, depleting stocks of such resources as fossil fuels, high-grade

minerals, fresh water (especially groundwater), forests, and fisheries. Instead, we must learn to live off the "interest" of the biosphere, so that the earth is able to sustain the human population through time. Given the realities of our current situation, how can we do this? Many believe that curtailing population growth is the only way to conserve the resources we need to survive; the addition of more and more people to the world each year may simply cancel out the advances the rest of us make in conservation and more efficient technology. In addition, those of us in developed countries use far more than our share of the earth's resources; we must learn to take advantage of more efficient technologies, recycle, and use only what we need.

Population growth and the problems of poverty and starvation which so often accompany it in less developed countries not only contribute to environmental degradation; they are also exacerbated by such degradation. When famine and poverty push people to overgraze fields, cultivate fragile soil, and cut down more trees than can be replaced, soil erosion and deforestation are only hastened. The prevalence of disease influences them to have many children in the hopes that a few will live. Thus, overpopulation, hunger, and abuse of environmental resources are tied into a vicious cycle that is very difficult to break.

The situation in Ethiopia, brought to international attention by the catastrophic famine of 1984-1985, is one of the most tragic examples of such a vicious cycle. Persistent drought, coupled with unchecked population growth, had exacerbated the problems of an agricultural system pushed to the brink of collapse by short-sighted, destructive farming practices. In addition, a history of oppressive governments, decades of civil war, and long-standing ethnic hostilities helped set the stage for disaster. Although countries and organizations all over the world rushed to give food and supplies to Ethiopia, the conditions that led to the famine persist and will continue to do so until a stabilizing federal government, acceptable to Ethiopia's various ethnic groups, makes a consistent, concerted effort to solve the root problems that make the nation especially vulnerable to food shortages.

PHYSICAL BOUNDARIES: THE CLIMATE AND SOIL OF ETHIOPIA

Ethiopia, one of the easternmost nations in sub-Saharan Africa, is bordered by the Red Sea to the northeast and also shares borders with the Sudan, Kenya, Somalia, and Djibouti. Most of north and central Ethiopia consists of high plateaus and mountains, with a strip of tropical lowland extending along the Sudan border.

One of the most important factors affecting Ethiopia's geography has been the action of Africa's Great Rift, one of the most extensive faults on the earth's surface. The rift runs 5,600 kilometers from the Red Sea south to Mozambique, with 480 kilometers passing through the heart of Ethiopia. The Rift is a spot where two plates underlying the earth's crust—the African Plate to the west and the Somali Plate to the east— have been gradually moving apart through the centuries. As they moved, the crust above them stretched and thinned, and hot magma (molten rock) welled up from beneath the plates, forcing the crust up into great domes. Eventually, the plates moved far enough apart to fracture the uplifted crust, creating Ethiopia's Great Rift Valley with the uplifted Western and Eastern Highlands to either side. The Western Highlands are characterized by massive mountains ranging from 8,000 to 12,000 feet above sea level, whereas the Eastern Highlands are lower, with more plateau regions. A less elevated geographic area includes the Rift Valley and the lowland plains in the south.

Despite recurring famines, Ethiopia is not without agricultural potential. In 1970, agricultural experts predicted that, with the use of improved farming practices, the country's climate and soil should be able to produce food for 100 million people—twice the country's population in 1987. Ethiopia's agricultural potential is tempered, however, by the fact that the nation's climate and soil possess unique characteristics that make erosion a constant threat and render some Western agricultural practices useless.

Climatic Conditions

Although Ethiopia lies just north of the equator, the temperature varies with elevation from extremely hot to nearly freezing. The lowlands and most of the Rift Valley are hot all year round. In contrast, the highest terrain experiences temperatures between 0° and 16° C (32° and 61° F). The highland plateaus are more temperate, with temperatures range between 16° and 29° C (61° and 84° F) annually.

Although a few areas of Ethiopia (primarily in the southwest) receive rainfall throughout the year, the weather follows a seasonal pattern in the bulk of the highlands. The rainy season usually begins with light rains in May and June, followed by a short period of hot, dry weather. Then, between June and September, torrential downpours dominate local weather. Normally, the Highland areas receive enough rainfall to make them Ethiopia's most agriculturally productive region. However, in recent years droughts have plagued the highlands, where they can be especially devastating because once a rainy season is missed, no precipitation can be expected for at least six months. When the rain does fall, it comes down with such force that any soil not protected by plant cover and secured by root systems is easily washed away, especially in the mountainous highlands.

Soil Characteristics

The soil itself poses unique agricultural problems because, in many areas, it contains an extremely high percentage of clay. This heavy soil is easily waterlogged during rainy spells. Consequently, farmers must either wait for the soil to dry out, shortening the growing period, or plant water-tolerant crops, which usually yield less grain than other varieties. As a result, they tend to plant crops on slopes, which drain faster than flat fields (these are often left for grazing). Unfortunately, cultivating the slopes makes them more vulnerable to erosion.

In fact, soil erosion has become a major problem in Ethiopia, and it is more serious than most farmers realize. Soil loss depresses yields for many years before the fields lose too much soil to be farmed. Half of the farmland in Ethiopia (35 million acres) has already been severely degraded by erosion, and 4.9 million acres have lost all their soil. Currently, the country loses about 3 billion tons of soil to erosion each year, and some soil experts predict that if this rate of erosion continues, the amount of acreage lost to agriculture will increase fivefold by the year 2010.

BIOLOGICAL BOUNDARIES: OVERUSED AND OVERLOOKED RESOURCES

Ethiopia is home to a diverse array of plant and animal species. Ironically, this country—associated with famine in most of our minds—is the historic source of a wide variety of grains. Some plant geneticists believe that varieties of millet, coffee, safflower, barley, and emmer (a type of hard, red wheat) were first domesticated in the Ethiopian area.

Some native grains possess great potential for alleviating hunger within Ethiopia—potential that foreign and domestic efforts alike have virtually ignored. For example, teff is an indigenous grain that can survive at higher altitudes than corn or sorghum, but also needs less rain than wheat or barley, the other principal high-altitude crops. Ethiopians in both rural and urban areas eat bread made from teff as a staple of their diet. However, farmers are rarely given teff seeds or encouraged to grow the native grain as a part of aid programs. Although teff possesses great genetic diversity, crop researchers have yet to try to breed even hardier, more productive strains that would benefit farmers even more.

Besides grains, Ethiopian farmers grow legumes, such as chickpeas or lentils; oilseeds, such as linseed and sesame; coffee; cotton; and sugarcane. Many farmers also raise livestock, primarily cattle. In fact, Ethiopia produces 25 percent of all the cattle raised on the African continent. Consequently, almost half of the arable land is used for grazing.

The highlands, though primarily grassland or agricultural fields, are dotted with such trees as the yellowwood and the juniper. Trees are more numerous in the lowlands, such as the savanna of the southern Rift Valley. The acacia is the most common tree in the lowlands. However, trees all over the country are threatened by massive deforestation: today less than 12 percent of the land sustains forest growth, compared to 40 percent in the late 1800s. Ethiopia's rapidly growing population has required more and more fuel and farmland as time goes by, and cutting down forests provides both. Unfortunately, unrestrained clearing of forests has almost exhausted the "crop" of firewood and left vast stretches of land unprotected against erosion. Some climatologists suggest that the clearing of forests might even have reduced the area's overall humidity, thus contributing to the persistent lack of rainfall that has plagued the region for over 20 years.

SOCIAL BOUNDARIES: THE POLITICAL AND CULTURAL ROOTS OF FOOD SHORTAGES AND POPULATION GROWTH

One of the least economically developed countries in the world, Ethiopia depends largely on agricultural production; over 80 percent of the population is involved in farming. Agricultural products, especially coffee, make up 90 percent of Ethiopia's exports, and account for 50 percent of its gross national product (GNP). How could a nation so oriented to producing crops be the victim of some of the worst famines the world has known? In part, the answer lies with the low yields of the average farmer—two-thirds of the country's farmers manage to produce only enough to stay alive. As a consequence, Ethiopia imports an average of 15 percent of its food requirements even in nonfamine years.

Ethiopia, like most other nations, depends upon its rural sector to subsidize the urban sector. (In the United States, government subsidies—which encourage farmers to maximize production—help to keep food supplies abundant and prices low.) The centralized Marxist government that prevailed in Ethiopa between 1977 and 1991 also wielded great influence over its farmers. Because the nation received aid from the World Bank, the government was compelled to show a financial return on the bank's investment. Consequently, it encouraged farmers to grow crops that could be sold to export markets at a profit, crops like coffee, strawberries, cotton, and carnations. These cash crops, as they are called, were grown in place of food crops that would otherwise be raised to meet local needs. Moreover, many peasant farmers were forced to work on state farms at the expense of their own small plots. Still others refused to grow more food than they personally needed because the government had set the prices for crops at artificially low levels.

Many observers see Ethiopia's rapidly growing population as the chief cause of its history of famine. Others maintain that Ethiopia has a potential carrying capacity much greater than its current population and should be able to feed itself for many years even with a rapidly growing population. However, the country's political system and agricultural practices have greatly limited its capacity to produce food, and thus a growing population simply causes food demand to outstrip production by a greater and greater margin every year. In years of major droughts, agricultural production drops even more, and even in years when it grows, it never does so by as large a percentage as the population. Therefore, Ethiopia must import more food every year.

Ethiopia's rapid population growth is due to an average birth rate that is roughly 2.5 times the death rate: between 1970 and 1981, 46 people were born for every 18.1 who died. Just between 1984 and 1990, the population grew from 46 million to almost 50 million, and this number would surely be higher if not for the soaring death rates during the 1984-1985 famine. Not only is population growing, but the rate at which it grows has been generally increasing. For example, population grew by an average of 1.8 percent each year between 1956 and 1965, but grew about 3.3 percent annually between 1980 and 1984. Although the growth rate slowed to 2.9 percent by 1989, some attribute this decline to the loss of many fertile men and women who either died or emigrated during the famine.

The World Bank estimates that fertility rate in Ethiopia, which stood at 6.1 in 1989, will not decline until 2045; thus, the population could grow as large as 231 million, over four times the present population. Based on the projection that the population will grow to 112 million by 2025, the United Nations predicts that Ethiopia will be able to feed only 36 percent of its people by the year 2000.

This destructive trend of unchecked population growth comes from several tendencies deeply embedded in the Ethiopian culture. First of all, the few efforts that the government and relief agencies have made toward family planning have been aimed only at women. Although women's participation is important, in a male-dominated culture like Ethiopia the men's understanding of the importance of family planning may be even more crucial. Most Ethiopian men consider fathering many children to be a sign of virility, and only extensive educational efforts and some sort of incentive will convince them to give up having large families. Currently, only 1.5 percent of couples report that they use any form of birth control at all.

Another factor that motivates Ethiopians to have many children is the likelihood of one or more children succumbing to starvation or illness. Poor or nonexistent health services and contaminated water contribute to an infant mortality rate of 155/1000, one of the world's highest. Multiple offspring ensure that parents will have enough children to help them work their farms. Thus, until better health care, clean water, and sufficient food help assure parents that the children they have will live, the incentives to produce as many children as possible will continue.

LOOKING BACK: THE HISTORY OF THE ETHIOPIAN REGION

Ethiopia traces its history to the ancient kingdom of Aksum, which came to be known as Abyssinia. Encompassing most of present-day northern Ethiopia, the empire was ruled by a dynasty thought to be descended from the legendary Menelik I, son of King Solomon and the Queen of Sheba. The long and complex political history of this country is intimately bound up with the forces that have brought about famine in the past and still threaten Ethiopians today.

Ethnic Groups and Class Structure in Ethiopia

During the late 1800s, Emperor Menelik II expanded his control by conquering provinces to the east, south, and west. By incorporating these areas into the empire, he extended its boundaries almost to their present day limits. Then, throughout the first half of the twentieth century, Ethiopia gained the Ogaden area along the Somali border through treaties with Britain, France, and Italy. Finally, in 1952, at the recommendation of the United Nations, Britain granted Ethiopia control of Eritrea, the area to the north of Ethiopia along the Red Sea. Formerly controlled by Italy, the people of Eritrea had hoped for nationhood after World War II.

Until its imperialist expansion of the nineteenth and twentieth centuries, Ethiopia was dominated by the Amharas, an ethnic group from the original, central provinces such as Shoa. Even today, the Amharas are at the top of the country's class structure. They are predominantly Christian, belonging to the Orthodox Ethiopian Church. However, the addition of many new provinces, each with their own cultures, religions, and loyalties, set the stage for ethnic and political unrest. Although the northern people of Eritrea and Tigre share the heritage of the Ethiopian Orthodox Church with the Amharans, they speak Tigrinya rather than Amharic. Eritrea in particular has never accepted rule by the Amharans of Ethiopia. When the Eritrean Parliament voted for full unification with Ethiopia in 1962, amid rumors of bribery and intimidation by the Amharan-dominated central government, Eritreans unhappy with the decision began a civil war that would last for three decades. Tigreans, also from the northern provinces, are ancient rivals of the Amharas; the animosity between the two groups dates back to the time when Aksum, located in the heart of Tigre, was the capital of the ancient Ethiopian kingdom.

Southern Ethiopia is home to a number of ethnic groups, many of which are Muslim. The largest of these groups, the Oromos, represents 40 percent of the Ethiopian population and has also manifested signs of desiring its independence from the Ethiopian state. In addition, many people in the southwest provinces of Hararge, Arussi, Bale, and Sidamo feel stronger ties with the people of Somalia, with whom they share an ethnic background, than with the dominant Amharans.

The Effect of Feudalism on Agricultural Methods

Farmers in Ethiopia originally practiced sustainable agricultural methods such as rotating crops, allowing land to lie fallow, and monitoring limited water supplies. In the twelfth century, however, the development of a feudal system—which gave control of the land and its products to nobility and landowners rather than the peasant farmers—caused farmers to gradually abandon these methods. Under Emperor Selassie, who ruled Ethiopia from 1930 to 1974, farmers comprised 90 percent of the labor force, but few owned their own land. Single landlords held more than two million acres of the country's arable land and received up to three quarters of what was produced on it. In the north, lands were generally managed by local gentry and feudal lords who collected a percentage of crop output as well as taxes. In the south, roughly three quarters of the land had been confiscated by the emperor, who kept it for himself or gave it as gifts to nobles or officers in the military. The peasants who farmed the land were obliged to give the landowners or managers money, free labor, provisions, and pack animals.

Farmers in neither the north nor the south had sufficient incentive to produce high yields or care for the land; abundant crops and healthy land merely provided extra profits for those already wealthy. When farmers needed to increase yields, they did so by abandoning crop rotation and by farming marginal land or areas that first had to be cleared of forests. Farmers frequently allowed livestock to overgraze fields. In addition, landowners encouraged the planting of cash crops, such as coffee, rather than food. Cash crops usually take an especially heavy toll on the soil's nutrients and benefit no one but the wealthy landowner.

This cycle of short-sighted overconsumption of the country's natural resources has made it more and more difficult for successive generations of farmers to supply the food needs of the country's growing population, further encouraging drastic means of increasing short-term yield. The "vicious circle" of destructive agricultural practices and population growth has made Ethiopia more and more vulnerable to the effects of droughts and other seasonal and cyclic weather variations.

The Seeds of a Revolution

In an effort to modernize Ethiopia, Emperor Haile Selassie proposed various ways to improve agricultural practices, but the landowning class made sure that they did not lose any of their privileges or land to the farmers. Legislation that would have directly helped farmers was never passed by the parliament, which was dominated by landowners. Programs that gave tools, seed, and livestock to peasants increased production for a time, but, without a change in legislation, landlords absorbed most of the benefits of the increases. The creation of modern farms equipped with Western technology favored foreign investors rather than peasant farmers.

As the gap between rich and poor in Ethiopia widened, a severe drought struck the country in 1972. The government refused to admit to the rest of the world that thousands of Ethiopians were starving, effectively preventing any food or money aid from entering the country. Although the continuing drought and famine became widely known by the end of 1973, starvation and misery had progressed far beyond the reach of foreign aid. The lack of water led to the misuse of wells. During the drought, so many people had to share the few available water sources that they became disease ridden. As a result, fewer than 3 percent of the population had access to uncontaminated water. Twenty million were perpetually hungry, and 8 percent of the population actually starved to death. The aid that did arrive was grossly mismanaged by the Imperial Ethiopian government, which used available motor vehicles to control the unsatisfied population rather than to transport food. The Ethiopian Orthodox church, which controlled a large proportion of the nation's arable land and thus owned supplies of livestock and grain, sold food to relief groups for as much as a 300 percent profit.

The rising discontent fueled a revolution in 1974, when the government was taken over by Soviet-backed Marxists. Within a few years, Lieutenant Colonel Mengistu Haile Miriam was firmly in place as the country's leader. However, the new regime failed to help the plight of the peasant farmer or to encourage soil and water conservation. The new government did nationalize land in Ethiopia, restricting the size of individual family farms to 10 hectares, but it did little to help farmers become self-sufficient. Nearly all of the funds spent on agriculture were directed to collectivized state farms; even so, these produce just 6 percent of the country's output. By far, the largest percentage of the government's money has been spent battling various internal rebel groups. To help feed the potentially rebellious cities and army, food prices were set approximately 70 percent below market level. World Bank monies were used to increase the production of cash crops rather than food staples. Unable to own land or profit from their labors, destitute peasant farmers have even less incentive than before the new regime to increase yields or take care of the land. Even so, these farmers, with little help from anyone, still produce 93 percent of the crops in Ethiopia.

The Famine of 1984-1985

The well-publicized famine of 1984-1985 was technically brought about by a drought that especially affected the northern provinces. An estimated one million people starved to death as a result. In 1984, the government began a "resettlement" program to move farmers from the drought-stricken north to the more fertile southern provinces. Farmers did not go voluntarily, but were tricked into government camps by promises of grain or cattle and then abducted and forcibly moved. Thousands of people were separated from their families, kept for days with very little food or water, and dumped in areas with no shelter, source of water, or agricultural equipment and forced to perform slave labor for the government. Many escaped to the Sudan, where they told stories of their inhumane treatment to foreign journalists. Human rights organizations estimate that 10 to 20 percent of the 500,000 Ethiopians resettled during 1984 and 1985 died in the process. Worldwide, many people suspected that Mengistu was trying to remove potential troublemakers from the rebellious northern provinces rather than help poor farmers.

Another practice criticized as inhumane was villagization. Claiming that it was easier to give farmers health, educational, and social services if they lived in centralized villages, the government sent out troops of soldiers to force farmers in the southern provinces to pack up their belongings and move. The homes of those who refused to go were burned. The farmers were often forced to move during agriculturally critical times, such as harvest or planting season, thereby forcing famine on what was previously one of the country's most productive regions. Villagization affected approximately ten times as many farmers as collectivization. As with collectivization, many victims fled the country, this time to Somalia. The subsequent lack of services provided to the "villages" confirmed many groups' suspicions that the government had again been trying to quell unrest. This program targeted another potentially rebellious group, the Muslim Oromos, who had been oppressed as a religious and ethnic underclass for centuries.

Foreign aid to minimize the damage inflicted by the drought was plentiful during this famine. By January 1985, the Ethiopian government had received $312.7 million and 509,916 metric tons of food from nongovernment international organizations alone. The United States, the largest donor, supplied 383,000 tons of food and $37.5 million in non-food aid.

Unfortunately, the Ethiopian government used this aid as a tool to suppress the discontented population of the northern provinces. Aid organizations were not permitted to provide food or money to anyone in the rebel provinces, even though the bulk of those starving were located there. Anyone receiving aid at a relief station was required to prove he or she belonged to a state-sponsored peasant organization. Government and army personnel have also been accused of confiscating equipment from relief workers, beating volunteers, and pillaging relief camps suspected of harboring rebels. The government raised money from the foreign aid by taxing shipment of relief grain and confiscated the grain and livestock that peasant farmers had managed to obtain in order to keep the army and urban centers content. Tins of donated butter oil were even found for sale in Ethiopian grocery stores.

As the famine went on, farmers went to greater and greater lengths to try to raise enough food to survive. Some even risked their lives to cultivate 90° slopes, working by hanging from harnesses high above the valley floors. Soil erosion and desertification grew increasingly worse, especially in the northern highlands, which were most affected and already stressed. When rain finally fell in the summer of 1985, the torrential downpour simply washed away soil, roads, bridges, and huts.

Rebels Overthrow a Dictator

Although Mengistu's government managed to survive the 1984-1985 famine, in the years that followed the Ethiopian people grew more and more discontented. A number of rebel groups grew in power, including the Eritrean People's Liberation Front, the Tigre People's Liberation Front, and the Oromo Liberation Front. Only two years after the 1984-1985 famine, another drought threatened to devastate the country. Although a full-scale crisis was averted by substantial foreign aid, civil strife continued to displace farmers, prevent food from reaching the hungry, and generally add to the misery. In 1988, the four most prominent rebel groups successfully united under the name of the Ethiopian People's Revolutionary Democratic Front (EPRDF).

When the rains failed again in 1989, international aid provided 650,000 tons of food and relief supplies. However, Mengistu refused to allow any of the aid to enter the northern rebel territories, where approximately 2 million people were in danger of starving by January 1990. He also announced that he would abandon proposed development plans for 1989-1990 in order to use those funds for the suppression of the rebels. The United Nations refused to provide any aid to the rebels against the government's wishes. Nevertheless, many individual countries and organizations did so, making the 800 mile overland trek from the Sudan to the northern provinces because the government still controlled the northern port of Massawa.

Meanwhile, as the EPRDF continued to gain in strength, Mengistu put more and more effort into trying to suppress the rebel group; by the late 1980s, he was spending approximately 70 percent of the country's revenues on weapons, most of which were used to fight the insurgents. In May of 1989, Mengistu made a desperate attempt to swell his army's forces; his soldiers kidnapped an estimated 50,000 teenage boys, who were sent with virtually no training to fight rebel forces. This incident combined with Mengistu's increasingly oppressive regime to add to the growing resentment of his government.

Finally, in May 1991, EPRDF forces, led primarily by Tigreans, came within 50 miles of Addis Ababa, the Ethiopian capital and Mengistu's stronghold. Deeming the war a lost cause, the Ethiopian army scattered and Mengistu fled to Zimbabwe. As the EPRDF, led by Meles Zenawi, took over Addis Ababa, allied forces of the Eritrean People's Liberation Front gained control of the government garrison in Asmara, the Eritrean capital, and of nearby Assab, a major seaport.

With the establishment of a provisional government by the EPRDF, Ethiopians were free of Mengistu, but a host of problems remained. The largely Amharan population of Addis Ababa appeared suspicious of the Tigrean-dominated EPRDF. The widespread fighting had closed roads and ports, preventing the transport of food; 8 million people were threatened with famine by June 1991. The Israeli government, which had been negotiating with Mengistu for the emigration of 14,000 Ethiopian Jews, took advantage of the confusion to airlift this group out of the country in just 33 hours. At last, the country took a step towards normalization on July 15, 1991, when 81 delegates representing 24 different groups adopted a charter that provides for multi-party elections in 1993.

Meanwhile, the Eritrean People's Liberation Front announced that it would administer the province until a vote on Eritrea's status is held, a vote that is expected by most political observers to result in secession from Ethiopia. The United States announced that it supported the Eritreans' right to self-determination. Many non-Eritreans oppose the province's independence, however, because without Eritrea's Red Sea coast, Ethiopia would be landlocked. Food aid and supplies, so critical in famine years, enter the country chiefly through the Eritrean ports of Massawa and Assab. The Eritreans have promised not to interfere with the flow of goods through their territory, but in a country rife with historic ethnic divisions, such promises are looked on with suspicion.

LOOKING AHEAD: THE FUTURE OF THE ETHIOPIAN REGION

Although Mengistu's government, with its political barriers to easing poverty and famine, has been overthrown, Ethiopia's future remains cloudy. The political situation in the country must stabilize before substantial reform can take place, and the new government must make a commitment to ensure the productive use of foreign aid, encourage sustainable agricultural methods, and promote the control of population growth.

Ethiopia's Changing Political Situation

The prospects for assembling a stable, enduring central government in Ethiopia appear slim. In the absence of the common enemy Mengistu, the military allies that helped to overthrow the former Marxist leader may turn to fighting among themselves. The E.P.R.D.F., dominated by the Tigreans, wants a unified country but has pledged not to oppose Eritrean independence. Meanwhile, many Oromos are also demanding a vote on either autonomy or independence for the southern provinces, a demand that has only heightened tensions between them and the Tigreans. Thus, the question of which provinces will form a new Ethiopia is unclear. What kind of government will eventually be established is also unclear. Although the Eritreans and the Tigreans claim to favor a market economy and political pluralism, both groups have formerly espoused Marxist philosophy and have had little experience with democracy.

Relief Efforts

As of 1992, famine and starvation still threatened some 7 million Ethiopians. However, the world has learned from past relief efforts that it is not enough to simply provide food in hopes of averting a crisis. The Ethiopian people must become self-sufficient by breaking the cycle of drought, unsustainable agriculture, and famine. Important steps that aid donors are taking in this direction include furnishing farmers with drought-resistant seeds and basic farming implements and providing villages with better water supplies and health services to prevent the spread of disease during droughts.

In addition, foreign efforts to help reform agricultural practices should adopt a new approach. In the past, the United Nations Food and Agriculture Organization (FAO) directed most of its help toward improving the technical aspects of agriculture. Unfortunately, most peasant farmers could not afford to adopt Western technological practices and lacked access to essentials such as spare tractor parts. In addition, many Western innovations did not translate well to the Ethiopian environment. For example, heavy equipment like tractors was markedly unsuccessful in the wet clay, sinking uselessly into the heavy soil. The Western moldboard plough simply carves the soil into hard, useless ridges of dried clay. Critics of the FAO point out that, with its large population, Ethiopia doesn't need labor-saving technology, but rather needs education in sustainable agriculture, research on practices that will work in Ethiopia's unique climate and soil, and essentials like seed and hand tools.

Some aid agencies have implemented successful food-for-work programs, using grain, cooking oil, and other products as incentives for farmers to practice methods that will conserve and even build up soil. For example, some farmers have been convinced to build terraces, which prevent soil from washing away on sloped fields. Others have planted eucalyptus and acacia trees on easily eroded slopes. Certain areas, such as the Ansokia Valley 200 miles north of Addis Ababa, have become green and fruitful as the result of such efforts.

Agricultural Programs

Although foreign agency efforts to improve Ethiopian agricultural practices are important, they will not be enough without the support of the Ethiopian government. Most farmers will not adopt conservation techniques—even with incentives—if they can't see advantages within a few years. The Ethiopian government needs to catalyze good agricultural practices by promoting research into practices that work with the native soil, climate, and grains. In addition, land reforms that would return ownership of farms to the farmers would give them a reason to care about their long-term effect on the land.

In addition, some groups suggest that an agricultural extension program be established. If their workers could visit farmers regularly, educating them about soil conservation and providing them with simple tools and seeds, environmentally sound methods would become more widespread. In some cases, people should be encouraged to leave overcrowded areas so that the carrying capacity of the land is not overwhelmed. However, these farmers must again be given incentives to move. Forcing them to a new area is inhumane, traumatizing the farmers and causing them to care even less about the effects of their actions.

Once farmers can be persuaded to make changes, they need to address a variety of problems. Overgrazing by Ethiopia's large herds of sheep, goats, and cattle limits land available for crops and encourages erosion. If the land they graze were used to grow crops that could be cut and fed to the animals, each acre of land could support five times as many animals. This more efficient method would free more land for other crops, eliminating the need to cultivate marginal land, and reduce erosion caused by overgrazing.

It is critical that the fertility of Ethiopia's soil, which has fared so badly over the years, is restored. This can be accomplished, in part, through reforestation, which will also protect land from erosion and produce firewood needed by the population. The native acacia tree, a nitrogen-fixer, could be planted on and around agricultural fields. These trees would not only add nitrogen, but also generate a 50-100 percent increase in soil organic matter, increase the capacity of the soil to hold water, and increase microbiological activity in the soil. Moreover, acacias would decrease wind erosion and produce fodder for livestock.

Another way to rebuild soil is to encourage composting, irrigation, and even no-till methods of agriculture. Some researchers in Ethiopia are experimenting with methods based on native practices that are uniquely adapted to Ethiopia's soil and climate. For instance, they have adapted a traditional plow so that it builds soil-catching terraces as the farmer plows; these terraces have been known to increase barley yields by 50 percent in two growing seasons. Another plow creates raised "broad" beds that allow water to drain more quickly. Still another innovative tool facilitates the no-till technique with a blade that runs beneath the surface of the field and cuts the roots of weeds without disturbing the soil. Other scientists are working on crops that would be uniquely adapted to growing in Ethiopia's soil.

Slowing Population Growth

Even if, as some people argue, political and agricultural reform could enable Ethiopia to support a much larger population than it currently holds, rapid growth must end sooner or later. Taking measures now will help ease the current food shortages while avoiding full-scale disaster in the future.

Many experts point out that farmers will remain opposed to family planning until the standard of living rises all over the country. People must feel sure that the children that they do have will live and that they have some economic security other than their offspring before they will see the merits of limiting family growth.

Education is another important factor in encouraging family planning. In a male dominated culture like that of Ethiopia, men's understanding of the importance of family planning is just as important as women's. Given the lack of a cohesive national government to sponsor an educational program, international groups propose establishing regional offices that will sponsor publicity programs and provide community health workers for each province or area. Regional and local health care centers would ensure that couples who want contraceptives or information about family planning have access to them.

Because so many of Ethiopia's interrelated problems grow out of its history, politics, and culture, we who live in the United States and other countries can have very little direct influence over the path that Ethiopia takes. However, we can influence the type of aid we are providing. Are we teaching Ethiopian farmers to be self-sufficient and conscious of soil conservation? Are we insisting that the government make appropriate reforms before we provide aid? Such small incentives may help the Ethiopian people to implement the changes needed to protect their environment, safeguard the integrity and self-sufficiency of the peasant farmers, and ultimately ensure the future health and welfare of the nation.

SUGGESTED READING

Beyer, Lisa. "Rebels Take Charge." *Time.* June 10 1991. 26-28.

Ehrlich, Paul R., and Anne H. "Population, Plenty, and Poverty." *National Geographic*. December 1988. 914-945.

Ethiopia: The Politics of Famine. New York: Freedom House, 1990.

"Falling Apart." *The Economist*. June 8 1991. 44-45.

Gilkes, Patrick. *The Dying Lion: Feudalism and Modernization in Ethiopia*. London: Julian Friedmann Publishers, Ltd., 1975.

Kaplan, Robert D. *Surrender or Starve: The Wars Behind the Famine*. Boulder, CO: Westview Press, 1988.

Keller, Edmond J. *Revolutionary Ethiopia: From Empire to People's Republic*. Bloomington, IN: Indiana University Press, 1988.

Korn, David A. *Ethiopia, the United States, and the Soviet Union*. Carbondale, IL: Southern Illinois University Press, 1986.

Levy, Bernard-Henri. "Worse than Hunger." *The New Republic*. December 15 1986. 13-14.

MacKenzie, Debora. "Can Ethiopia Be Saved?" *New Scientist*. September 24 1987. 54-58.

————. "Ethiopia's Grains of Hope." *New Scientist*. August 20 1987. 20-21.

————. "Ethiopia's Hand to the Plough." *New Scientist*. October 1 1987. 52-55.

Mengisteab, Kidane. *Ethiopia: Failure of Land Reform and Agricultural Crisis*. New York: Greenwood Press, 1990.

Ottaway, Marina, ed. *The Political Economy of Ethiopia*. New York: Praeger, 1990.

Schwab, Peter. "Political Change and Famine in Ethiopia." *Current History*. May 1985. 221+.

Shepherd, Jack. *The Politics of Starvation*. Washington: Carnegie Endowment for International Peace, 1975.

Stager, Curt. "Africa's Great Rift." *National Geographic*. May 1990. 2-42.

Varnis, Steven L. "The UN's FAO: Is it DOA?" *Society*. September/October 1988. 38-41.

Walker, Brian K. "Famine in Africa—The Real Causes and Possible Solutions." *Environmentalist*. August 1985. 167-70.

CHAPTER 7

FOOD RESOURCES, HUNGER, AND POVERTY: HUNGER IN THE UNITED STATES

For most of us in the United States, the problem of hunger is confined to what we learn about famines in far-off countries on the news and what we see of starving children in magazine photos. But how many of us are aware of the true scope of world hunger, or of its causes and effects? Worldwide, about one in every five people suffers from chronic hunger, more than one billion people in all. Children account for about 40 percent of the world's hungry; the majority of the rest are women. Every year, an estimated 13-18 million people die from starvation and hunger; counting deaths due to hunger-related diseases as well, the figure is 30-40 million. Each day, hunger and starvation kill 35,000 human beings; every two days, the number of people who die from these causes is equal to the number who were killed instantly when the atomic bomb exploded on Hiroshima.

Most of the world's hungry live within the "Great Hunger Belt," a region that encompasses various nations in Southeast Asia, the Indian subcontinent, the Middle East, Africa, and the equatorial region of Latin America. Just five nations—India, Bangladesh, Nigeria, Pakistan, and Indonesia—are home to at least half of the world's hungry.

As we might suspect, environmental degradation such as soil erosion contributes to poverty and hunger. Although much of the world's environmental degradation is due to the industrialized, high-consumption lifestyles of people in developed countries, poverty, especially in developing countries, contributes to environmental degradation as well. A hungry family's concern for long-term conservation is far outweighed by the immediate need to survive. Lacking arable lands or decent paying jobs, they must try to squeeze out a living on marginal lands where soils cannot sustain cultivation, overgraze grasslands and pastures, or clear forests for fuel, food, or marketable resources. For example, forty percent of Guatemala's arable land has been lost to erosion. In Haiti, practically no topsoil remains for cultivation. In Bangladesh, Nepal, and Java, severe pressure on hillsides for fuel and growing crops has caused much erosion and flooding in lowland areas. Worsened environmental conditions, in turn, have led to more severe poverty and hunger.

Although environmental degradation is locked into this vicious cycle with hunger and poverty throughout much of the world, many people believe the cycle is sustained by underlying cultural factors such as social practices, politics, and economics. Hunger exists not

72

because we lack the technical and scientific know-how to eradicate it; hunger exists because people lack the political and economic power to fight it. One major example of the way politics contributes to poverty and hunger is the issue of land ownership. The political machinery of many developing nations prevents most people from owning the land they cultivate; as a consequence, farmers have little incentive to preserve soil or soil quality.

Despite the prevalence of hunger in the world, other people are far from hungry. In many developing nations, there is a widening gap between the privileged class and the majority of people; the wealthiest 20 percent of a country is often ten to twenty times richer than the poorest 20 percent. In the United States, people spend billions of dollars each year on special diets in the hope that they can "control" their appetites and "shed unwanted pounds," while 500 million people worldwide are so undernourished that their bodies and minds are wasting away. Even within the United States, a significant proportion of people experience hunger; experts estimate that at least 20,000 Americans go hungry for at least some time during each month, and some are hungry every day.

Hunger in the United States is not linked to environmental degradation in the direct way it is in many developing countries; although the United States produces more than enough food to support its population, the American hungry simply cannot afford to buy what they need from this vast supply. However, the hunger problem in the United States is representative of global hunger in that political and social barriers prevent the hungry from receiving the assistance they need.

PHYSICAL AND BIOLOGICAL BOUNDARIES:
THE VICTIMS AND THE CAUSES OF HUNGER

Hunger hurts everyone who experiences it, but children and the elderly are at the greatest risk of sustaining serious medical complications. Children need the proper nutrition and caloric intake to grow and develop both physically and mentally. Severe protein insufficiency may cause mental retardation, physical wasting away, and death. Insufficient protein also makes children vulnerable to a number of diseases. A study conducted by Maine's Department of Human Services showed that children of poor families were three times as likely to die as other children, primarily from disease. Bread for the World, a nonprofit agency working against hunger, estimates that 27 U.S. children die every day from the health-related effects of poverty, often those linked to inadequate diets. Teachers report that children who come to school hungry have a hard time concentrating and learning.

Children of poor families are at risk even while they are still in the womb. If their mothers do not eat sufficient amounts of nutritionally balanced food, the fetuses cannot develop. This trend is one factor which contributes to a high infant mortality rate. In fact, as of 1989, the U.S. rate was higher than that of 18 other industrialized countries, including Japan, Belgium, and the United Kingdom.

Many elderly people need medically prescribed diets to stay healthy. Eighty-five percent of people over 65 suffer from chronic conditions that influence their nutritional needs. Others are unable to absorb nutrients properly and need special, highly nutritious foods. In addition, as people grow older, they usually become more susceptible to illness, and lack of nutrition exacerbates this vulnerability. Unfortunately, the millions of elderly who live below the poverty level are unlikely to be able to afford special diets when they need them. Those ineligible for food stamps are at even higher risk of illness and death.

Because hunger is a "silent epidemic," it is hard to pinpoint exactly who is hungry in America. However, we know that people who live in poverty are most likely to be hungry. The U.S. government defines poverty as living on less than the income level necessary to guarantee health and survival; as of 1989, the poverty line was $12,091 for a family of four and $6,017 for one person. In 1987, roughly 32.5 million Americans were living below these income levels. However, many concerned citizens, including groups like Second Harvest (a nonprofit national food bank network), see poverty as being unable to attain what the majority of the population would consider an adequate standard of living. No one knows how many additional people would qualify as poverty-stricken under this definition.

Some people wonder why the poor don't simply get jobs. In fact, many poor people do work; as of 1989, more than 5.5 million people living below the poverty level belonged to households where someone worked full-time year-round. Unfortunately, one adult working full-time for minimum wage cannot earn enough to bring his or her family out of poverty. In addition, over half of the poor are unable to work; approximately 40 percent of them are children, and 10 percent of them are over 65. Many of the non-elderly adult poor cannot work for a wide variety of reasons, including disability, illness, mental problems, lack of education, or simply a lack of jobs in their areas. Some who can work must stay home with their children because 52 percent of poor families are headed by a single parent.

SOCIAL BOUNDARIES: HOW HUNGER INVOLVES EVERYONE

The U.S. government has a variety of programs to help the poor avoid hunger. Nonetheless, 20 million people in this country are still going hungry for some time during each month for a variety of reasons. Many families who receive food stamps are unable to make them last the whole month, and thus go hungry the last few days or more. Other, recently unemployed people usually have assets like cars or homes which they are unwilling to give up in order to get food aid. Others were "weeded out" of federal aid programs when the Reagan administration made substantial budget cuts in the early 1980s. The growing population of homeless people cannot obtain federal food aid because they have no fixed addresses.

The prevailing government attitude toward hunger in the 1980s was that as the economy revived, poverty would disappear, and so would hunger. In the meantime, the hungry simply had to apply for government aid, learn to budget and eat nutritional food, and abstain from alcohol and drugs. Although the economy did improve throughout the mid-1980s, poverty did not disappear. According to the U.S. Bureau of Census, by 1985 the wealthiest two-fifths of the U.S. population received 67.3 percent of the national income, while the poorest two-fifths received only 15.7 percent. This gap between rich and poor was the widest the Census Bureau had ever recorded since it began gathering income information in 1947. As of 1987, this gap had even grown slightly wider.

One reason that the poor became still poorer in the 1980s was the effect of the budget and tax acts of 1981 and 1982. Between 1983 and 1985, these acts took $23 billion from families with annual incomes under $10,000 and contributed $34.9 billion to those with over $80,000, according to research carried out by the Physician Task Force on Hunger (PTF). A study conducted by the nonpartisan Congressional Budget Office further confirms that the distribution of federal benefits and tax breaks works to widen the gap between rich and poor. By merging data from both the Census Bureau and the Internal Revenue Service, they found that, in 1991, households making over $100,000 per year received an average of $5,690 in federal cash and benefits, while households making under $10,000 per year received only $5,560. The big picture is even more telling. From 1980 to 1991, households with annual incomes under $10,000 lost an average of 7 percent of their federal benefits when the amount was adjusted for inflation. However, for households with annual incomes over $200,000, the value of benefits received (chiefly Social Security, Medicare, and federal pensions) actually doubled in the same eleven year period.

Many better off citizens feel that federal programs which aid the poor and hungry are a drain on their tax money. But by ensuring that everyone has enough to eat, we avoid the economic and social costs associated with malnutrition, illness, and unemployment. For instance, when pregnant, low-income women participate in government food supplement programs, they have 23 percent fewer infants with low birth weights and fewer premature births and infant deaths. Children who participate in these programs avoid mental retardation and illness and are better able to develop social and work skills. In addition, those who suspect that the food stamp program is abused by people who should not qualify would be surprised to learn that over 95 percent of households receiving food stamps *are* eligible—a higher percentage than for those receiving income tax refunds, according to the Physician Task Force on Hunger.

LOOKING BACK: THE HISTORY OF HUNGER IN THE UNITED STATES

The federal government first became involved in fighting hunger in 1939, when it began a food stamp program. However, this program was aimed more at distributing farm surpluses than feeding the hungry, and was curtailed in 1943 because of food shortages linked with World War II. The surprisingly high number of young men rejected from the World War II draft because of nutritional deficiencies spurred Congress to initiate the National School Lunch Program in 1966. By the 1960s, approximately one-fifth of the U.S. population was living in poverty and at risk for hunger. At that time, the government established or revised many of the assistance programs that are with us today.

How Federal Programs Help Combat Hunger

The Federal Food Stamp Program, the largest federally subsidized program, helps supplement the food-purchasing power of the poor. This program is especially important because it can be used by the working poor as well as the unemployed. Stamps can be exchanged for food at grocery stores but cannot be used to purchase alcohol, tobacco, imported food, or household products. Each household receives enough stamps so that, combined with a third of its income, it can afford the Thrifty Food Plan devised by the U.S. Department of Agriculture. For example, in 1987, the maximum allotment (given to households with no income) for a family of four was $268. As of 1986, 19 million people received food stamp benefits.

School breakfasts and lunches are major programs that help alleviate hunger among children. The government provides subsidies to public and private schools and day care centers so they can provide meals for the students. Although some of these children pay for their meals, those from poor families receive them for free or reduced prices. As of 1987, 23 million children participated in the lunch program, 3.3 million in the breakfast program, and 1 million in the child care food program.

Another important form of food assistance is the special supplemental feeding program for women, infants, and children, commonly known as WIC. WIC serves pregnant and post-partum women, infants, and children up to five years old who live in low-income households. Participants receive vouchers for foods high in protein, iron and calcium so that inadequate diets do not endanger their health. As of 1988, the program served over 3.5 million; owing to insufficient funds, another four million hungry were not served.

Other forms of assistance include the Temporary Emergency Food Assistance Program (TEFAP) and elderly nutrition programs, which help fund programs like group meals and delivering meals to the housebound. Finally, Aid to Families with Dependent Children (AFDC), although it does not provide direct food assistance, gives poor families money that they may use to purchase food. However, in about half of the states, families cannot receive AFDC if both parents are at home. As of 1984, 10.9 million people (60 percent of them children) received AFDC assistance.

How Hunger Reappeared in the 1980s

Although these and other assistance programs helped reduce hunger in the United States by the end of the 1970s, the trend turned around in the 1980s. Concerned by the mounting federal budget, the Reagan administration cut $13 billion from food assistance programs between 1982 and 1985.

These cuts eliminated nearly one-million people from the food stamp program, and reduced benefits to many others; by 1987, fewer people were participating than in 1980 even though over 6 million more had fallen below the poverty level in that time. Hundreds of thousands of families no longer received AFDC. Three million children were cut from school lunch programs and 500,000 from breakfast programs. Many summer food programs for school children lost funding. Although President Reagan claimed that these cuts in food assistance would not hurt the truly needy, two-thirds of spending reductions came from cuts in aid to families below the poverty level.

The effects of the budget cuts began to show at private, nonprofit emergency food programs like soup kitchens and food banks. The Center for Budget and Policy Priorities reported that one-third of the country's emergency food organizations received 100 percent to 200 percent more requests for help in 1983 than they had in 1982. In 1984, the Food Research and Action Center surveyed 300 emergency food programs across the country and discovered that almost 75 percent believed that the private charities in their community could not meet local needs.

In response to the growing criticism of the way the budget cuts had increased hunger, President Reagan appointed a task force to study the problem. Although the task force's final report claimed that they could neither refute nor support claims of widespread hunger, they did make some recommendations designed to combat what hunger did exist. These included simplifying food stamp application procedures and allowing individuals without fixed addresses to participate.

Many people criticized the presidential task force for skirting the issue of hunger. In fact, many other groups, including government agencies, universities, and nonprofit groups, conducted official studies on hunger in the early 1980s, and they all agreed that hunger was a serious and growing problem in the United States.

One of the most thorough of these studies was a report by the Physician Task Force on Hunger in America, a group of 22 physicians under the leadership of J. Larry Brown, a physician with the Harvard School of Public Health. Throughout 1984, the physicians visited cities and towns in the South, Northeast, Midwest, and Southwest to find out how widespread the hunger problem was and why it existed. Their conclusion was that "the problem of hunger in the U.S. is now more widespread and serious than at any time in the last ten to fifteen years. It has returned . . . because the programs which virtually ended hunger in the last decade [the 1970s] have been weakened."

Groups concerned with hunger documented not only the statistics of hunger in the 1980s, but also individual stories of suffering. For instance, in 1983 a Los Angeles daycare teacher told representatives from a hunger action group that her center had lost funding for lunches. Since then, she said, the children would run out to the parking lot during lunchtime and fight over food from the garbage cans.

Not only did the budget cuts eliminate people who had nowhere else to turn, but they also curtailed outreach programs which helped people understand and obtain food stamp benefits. Outreach workers used to help people fill out forms and obtain proof of eligibility, and they even transported the elderly and disabled to the food stamp office. Eliminating this service prevented many who were eligible from obtaining their benefits. On the other hand, some employees of the food stamp program point out that new requirements established by the cuts require a substantial amount of extra paperwork, consuming time and money that would be better spent on hungry people.

Besides condemning the cutbacks in food assistance, the Physician Task Force (PTF) pointed out basic problems in the federal programs which caused unnecessary hunger and misery. Many of the programs' functions are carried out with little regard for the feelings of the human beings involved. For example, some physicians talked with an 83-year old woman in Montgomery, Alabama, who was blind in one eye. She told them her food stamps had been cut off, but no one had even told her why. In another case, the father of a family on food stamps reported that he had earned $22 for cleaning a lot. Although he reported this sum, he and his family lost their food stamps for a month because he did not have a receipt.

The physicians also talked with a young father in a soup kitchen in Montgomery. He had been laid off from his job at a steel mill, and after his unemployment ran out, he had been forced to leave home so his family could receive AFDC benefits. Looking for work, he wandered from town to town with no place to sleep. Although he was very lonely for his family, he couldn't even call them because they had no phone.

One major flaw in the food stamp program is the long periods of time applicants must wait to find out if they are eligible. The PTF noted that it was not unusual for people in St. Louis and Kansas City to wait 45-60 days for a decision on their eligibility. They also described a Mississippi family of 13 who were waiting to hear about eligibility. The family lived in the remains of a house with no windows, heat, electricity, food, or milk for the babies. Long waits sometimes occur because food stamp offices often neglect to ask applicants if they have anything to live on during the waiting period or to inform them of expedited issuance stamps.

Many food stamp offices prominently post signs warning applicants that they will be prosecuted for falsifying eligibility, which creates an intimidating atmosphere of suspicion that can discourage applicants. A food stamp administrator in North Carolina told the PTF about an elderly woman who finally came in to apply, but was trembling as she filled out the forms. It turned out that she was afraid of making an error and being jailed for fraud.

The sheer length of food stamp applications also deters potential participants. In Mississippi, an applicant must fill out 21 forms totaling 35 pages to apply for food stamps. In addition, many forms are too complex for the average applicant to understand. A readability analysis of the Minnesota application form showed that a reader would need to be in approximately the second year of graduate school to comprehend it.

In addition, the PTF called the nutritional value of the Thrifty Food Plan into question. They claimed that the USDA's primary goal was to devise a food plan to match a predetermined expenditure level, not to determine an adequately filling and nutritious diet.

How Private Organizations Help

Increasingly, private organizations are being called upon to assist the growing number of hungry falling through the holes in the federal government's "safety net" of assistance. In 1987, the Northern California Hunger Action Coalition noticed that food stamp offices were actually recommending that applicants go to private emergency food agencies while they waited for eligibility or instead of applying at all.

The two major types of private charities distributing food directly to the poor are food pantries and soup kitchens. Food pantries are small-scale operations that distribute grocery items to needy individuals. Soup kitchens provide hot meals directly to the hungry. Although some of these organizations require applicants to meet eligibility standards, most try to provide food to anyone who asks. Unfortunately, food pantries and soup kitchens frequently do not have enough food for everyone who comes to them.

Food banks collect food, often in the form of surpluses from corporations and retailers, and distribute it to food pantries. Many food banks use gleaning programs, where volunteers go into fields, orchards, and packaging plants to collect food that is left unharvested or cannot be sold because it is slightly bruised, too large or too small, or not attractive enough.

Because they are nonprofit, most food banks rely heavily on volunteers. For example, the Westside Food Bank in Surprise, Arizona distributes food to 100 charities that, in 1987, served a total of 15,000 families. Although Westside has a paid staff of only four or five, its four to five thousand volunteer gleaners and 600 other volunteers make its success possible.

One major problem food banks face is not having enough vehicles or personnel to transport food items offered to them. For instance, the Potato Project, which distributes potatoes to food centers in 47 states, couldn't afford to ship 10 million pounds of the potatoes it had gathered in 1987. Food banks also frequently suffer from insufficient freezer, refrigerator, or storage space.

Second Harvest, the largest nongovernmental food distribution program in the United States, is an umbrella organization for food banks across the country. It distributes to 88 food banks and 112 of their affiliates in 46 states. These banks serve over 38,000 private charities, including day care and senior centers, drug and alcohol treatment centers, and homeless shelters as well as food pantries and soup kitchens. Incredibly efficient, Second Harvest distributes $157 worth of food for every dollar it spends.

Second Harvest receives products from major corporation such as Beatrice Foods, Kraft, Pillsbury and General Mills as well as grocery stores and other retailers. In addition to distributing all sorts of food, including cookies, quiche, and canned vegetables, they sometime offer products like fabric softener, shoes, soap, and even stuffed bears. Second Harvest attracts powerful donors because it requires member food banks to meet strict standards for warehouse management, record keeping, distribution, and refrigerator and freezer capacity.

Other groups do not gather or distribute food, but concentrate on strengthening government programs that help the poor and hungry. One of the largest of these is the Food Research and Action Center (FRAC). FRAC's efforts include documenting hunger in the United States, monitoring public policy, keeping the public informed, working with concerned individuals and groups, and providing legal assistance on food aid issues.

How Recent Legislation Has Aided the Hungry

In the late 1980s, the widespread hunger in the United States finally made an impression on the government. Although President Reagan threatened to veto the Hunger Prevention Act passed by Congress in 1988, he eventually signed it. This act provided $3 billion between 1988 and 1993 to strengthen domestic food assistance programs. As a result of these funds, the average food stamp recipient received a real (adjusted for inflation) increase of 1.5 cents per day in 1989, 5 cents in 1990, and 8 cents in 1991.

In 1989, Congress also took a small step to help bring the working poor out of poverty. They voted to raise the minimum wage to $4.25 an hour by 1991. They also increased funding for WIC by $118 million, allowing the program to serve 200,000 additional people in 1990.

LOOKING AHEAD: THE FUTURE OF THE FIGHT AGAINST HUNGER IN THE UNITED STATES

Despite recent legislation aiding the hungry, many believe that government action alone will not end hunger in the United States. In fact, shortly after President George Bush took office in 1989, he urged private citizens and charitable groups to assume a greater share of the burden in solving social problems like homelessness and hunger. Others maintain that, although private charities can help provide food, they cannot lift families out of poverty. Instead, many people contend that we can end hunger only if we are willing to change the values and structures of our society.

The National Council of Churches published a Hunger Action Agenda in 1987, stating that

> Concern for the hungry must now lead to a critical look at the structural arrangements in our society for distributing wealth, income, power, work and status and to the underlying values and assumptions that have generated these arrangements and have kept them operative. Our society's assumption has been that competitive pursuit of

private gain will work to provide the best possible life for all. Our examination of hunger in the United States convinces us that it has not provided the best life possible for all, but has institutionalized the values of greed, acquisitiveness, competition, uncontrolled growth, excessive individualism and militarism, and has resulted in the exploitation of many for the sake of the few.

They not only recommend specific changes in legislation and programs, but also suggest ways that individuals can make a difference, such as organizing advocacy groups on specific issues.

The Institute for Food and Development Policy, also known as Food First, is a nonprofit research and education organization dedicated to investigating and exposing the root causes of hunger worldwide, including the United States. They maintain that hunger is not caused by scarcity, but by imbalances of economic and political power that keep food from those who need it. They provide a variety of educational publications and encourage political involvement to develop policies that will combat hunger.

Bread for the World is a Christian citizens' group that concentrates on lobbying for actions aimed at alleviating hunger worldwide. They advocate that the United States cut military spending by $20 billion (out of the $300 billion we spend currently) and use half of that for programs which combat hunger and its causes. They point out that even defense experts like Robert McNamara, former U.S. Secretary of Defense, say we can reduce military spending by 50 percent over the next 6 to 8 years without endangering national security.

Most concerned groups agree that we cannot end hunger simply by giving food and money to the hungry. They suggest that we must make ending hunger a true priority by establishing every citizen's right to a decent lifestyle, including enough to eat. Voting allows U.S. citizens have some say in determining their country's politics. However, to end hunger and poverty, we need to extend our democratic rights to economic issues, too. Food First points out that economic decisions in our country are made by corporations, for the short-term benefit of corporations. Even legislators who vote on economic issues often rely on corporate contributions. Should they be allowed to waste billions on mergers rather than creating jobs? Shouldn't we the citizens have more say in decisions like whether we should spend money on defense systems like the Stealth bomber, which may or may not work, or on programs that fight hunger and clean up pollution? We must realize that our "free market" economy, which prioritizes only profit, should also guarantee everyone the right to a decent job with sufficient pay and make provisions for those who cannot find jobs or cannot work. Only then will the United States be a true democracy, providing all its citizens the right to life, liberty, and the pursuit of happiness.

SUGGESTED READING

Brown, Larry J. "Hunger in the United States." *Scientific American*. February 1987. 36-41.

Cohen, Barbara E. "Food Security and Hunger Policy for the 1990s." *Nutrition Today*. July/August 1990. 23-27.

Cross, Audrey Title. "Politics, Poverty, and Nutrition." *Journal of the American Dietetic Association*. August 1987. 1007-1010.

Griffin, Keith. *World Hunger and the World Economy and Other Essays in Developmental Economics*. New York: Holmes and Meier, 1987.

Howe, Neil, and Phillip Longman. "The Next New Deal." *The Atlantic*. April 1992. 88-99.

Hunger and Society. Ithaca, NY: Cornell University, Program in International Nutrition, 1988.

The Hunger Project. *Ending Hunger: An Idea Whose Time Has Come*. New York: Praeger, 1985.

Kutzner, Patricia L. *World Hunger: A Reference Handbook*. Santa Barbara, CA: ABC-Clio, 1991.

Lappe, Frances Moore. *Diet for a Small Planet*. New York: Ballantine Books, 1982.

——, and Joseph Collins, with Cary Fowler. *Food First: Beyond the Myth of Scarcity*. New York: Ballantine Books, 1979.

Levitan, Sar A. *Programs in Aid of the Poor*, 5th edition. Baltimore: Johns Hopkins University Press, 1985.

Leinwand, Gerald. *Hunger and Malnutrition in America*. New York: Franklin Watts, 1985.

Mayer, Jean. "Hunger and Undernutrition in the U.S." *Journal of Nutrition*. August 1990. 919-923.

O'Neill, Onora. *Faces of Hunger: An Essay on Poverty, Justice, and Development*. London: Allen and Unwin, 1986.

President's Task Force on Food Assistance, *Report of the President's Task Force on Food Assistance*. January 1984.

Physician Task Force on Hunger in America. *Hunger in America: The Growing Epidemic*. Middletown, CT: Wesleyan University Press, 1985.

"Position of the American Dietetic Association: Domestic Hunger and Inadequate Access to Food. " *Journal of the American Dietetic Society*. October 1990. 1437-1441.

Schwartz-Nobel, Loretta. *Starving in the Shadow of Plenty*. New York: Putnam, 1981.

For more information on hunger in the United States and how you can help, contact the following organizations:

Bread for the World
802 Rhode Island Avenue NE
Washington, DC 20018
202-269-0220

Center for Budget and Policy Priorities
236 Massachusetts Avenue, NE Suite 305
Washington, DC 20002

202-544-0591

Children's Defense Fund
122 C Street, NW
Washington, DC 20001
202-628-8787

Ecumenical Child Care Network
Child Advocacy Office
475 Riverside Drive, Room 572
New York, NY 10115
212-870-3342

Food Research and Action Center
1319 F Street, NW #500
Washington, DC 20004
202-393-5060

Interfaith Action for Economic Justice
110 Maryland Avenue NE
Washington, DC 20002-5694
202-543-2800

National Council of Churches
Washington Office
110 Maryland Avenue, NE
Washington, DC 20002
202-544-2350

Physician Task Force on Hunger in America
Harvard University School of Public Health
677 Huntington Avenue
Boston, Massachusetts 02115
617-732-1000

Second Harvest
National Food Bank Network
343 South Dearborn Street
Chicago, Illinois 60604
312-341-1303

CHAPTER 8

ENERGY ISSUES: RECOVERY AND TRANSPORTATION OF ALASKAN OIL

Our society's demand for energy lies at the root of many of the most crucial environmental issues of the twentieth century. It is our demand for electricity that spurs the recovery and burning of coal, and thus causes the social and environmental tragedy of strip-mining and the air-borne plague of acid precipitation. It is our reliance on petroleum in order to run our automobiles and to manufacture artificial fertilizers and pesticides that leads to global warming, oil spills, and even the widespread soil erosion associated with high-input, petroleum dependent agriculture.

This need for energy is not unique to human beings. All living organisms require energy to grow, maintain life processes, and reproduce. Unlike other species, however, humans—at least those who live in industrialized societies like the United States—use up energy resources in amounts far beyond what is necessary for the ecological survival of our species. In addition, we are exhausting our available resources at a rapid rate. Most of the energy sources we use, such as the fossil fuels oil and coal, are classified as nonrenewable resources; they exist in finite supply or are renewed at a rate slower than that of consumption. Renewable resources, which are resupplied at rates greater than or consistent with use include solar, biomass (wood, plant residues, and animal wastes), wind, geothermal, and tidal. Although renewables cannot be exhausted (except for extreme cases, such as forests being cut down at such rapid rates that they cannot regenerate), most industrialized nations use them only to a limited extent.

The demand for energy has increased dramatically as the societies of industrialized countries have become more technologically complex. Most of the world's consumption of fossil fuels like coal and oil —from the dawn of recorded history until the start of the 1990s—has occurred in the last two to three decades. World energy demand rose almost 600 percent between 1900 and 1965 and is expected to increase another 450 percent between 1965 and 2000.

For the most part, we in the developed world are indifferent to the origins of the energy we use. Except for a brief period in the 1970s, when the Organization of Petroleum Exporting Countries (OPEC) embargo disrupted oil supplies, an inexpensive and plentiful supply of energy has always been available. We have largely taken this energy for granted, and ignorance of our energy sources has caused us to make several dangerous assumptions. Chief among these is the assumption that energy will always be available at prices we can easily afford. Moreover, because we are generally unaware of the sources of our energy supplies or how those sources are located, extracted, and transformed, we are typically unaware of the environmental effects of energy exploration, extraction, conversion, and use.

In the years immediately ahead, we—as individuals and as a nation—will make many choices concerning energy consumption and conservation. Each choice will offer benefits, and each will exact environmental, social, economic, and political costs. One situation that dramatically illustrates the issues involved in our society's patterns of energy consumption is the drilling and transporting of Alaskan oil. Alaska, the nation's largest state, contains more acreage than Texas, Montana, and California combined, and its relatively unspoiled stretches of wilderness areas provide refuge for many animals and plants. The state's human population is relatively small, and a significant proportion of the people outside its few major cities are Native Americans who depend on the land for survival. However, millions of other Americans have come to rely on Alaska's North Slope because of the oil that lies below its surface, and the needs and desires of these three groups have often come into conflict.

PHYSICAL BOUNDARIES: THE PRESERVATION OF A DELICATE BALANCE

Geographically, Alaska can be divided into four main regions. The southernmost region, the Pacific Mountain System, encompasses a number of mountain ranges, including Mt. McKinley, the highest point in North America. This area is bordered by the Aleutian Islands and Alaska Peninsula to the southwest, the Alaska Panhandle to the southeast, and the Alaska Range to the north. The Central Uplands and Lowlands, which begin above the Alaska Range, are characterized by low, rolling hills and the state's longest river, the Yukon. Further north rises the Brooks Mountain Range and its foothills, and the North Slope, also known as the Arctic Coastal Plain, extends from the Brooks Range to the Arctic Ocean. Potentially vast reserves of petroleum lie deep below the frozen North Slope tundra and offshore beneath the waters of the Arctic Ocean. Those oil reserves may well determine the fate of the Alaskan wilderness.

The tundra of the North Slope is land covered with permafrost, a permanently frozen layer of earth approximately 300 meters thick. Although air temperatures in northern Alaska fluctuate from 21° C to -21° C (70° F to -70° F) with the seasons of the year, the bulk of the permafrost layer always remains frozen because of its unique composition: frozen muck and gravel, covered by an insulating layer of soil, moss and grass. If this overlying layer is stripped from the top, the entire permafrost will melt when temperatures rise. Temperatures significantly higher than 21° C (70° F), such as those emitted by hot oil or drilling rigs, can melt the permafrost layer even without stripping away the insulation. Once permafrost melts, it may never regain the delicate environmental balance of life-supporting soil above and frozen foundation below.

Approximately 1.5 to 3 kilometers beneath the permafrost lies the oil that has brought the North Slope into national prominence. The Atlantic Richfield Company first discovered oil in the North Slope in April 1968, at Prudhoe Bay. Original estimates predicted that this area could produce 96 billion barrels of oil, enough to meet approximately 5 percent of the U.S. demand for 13 years. As of 1991, the area had produced approximately 8 billion barrels, with yearly production expected to decline beginning at the turn of the century.

BIOLOGICAL BOUNDARIES: ANIMALS
AND PLANTS OF AN EXTREME CLIMATE

Although Alaska is famous for its long, bitterly cold winters, climates vary throughout the state. Even in the coldest, most harsh Alaskan regions, however, many species of plants and wildlife manage to thrive. In June, when the sun begins its period of continuous daylight, the surface of the permafrost thaws enough that the North Slope is covered with a wide variety of flowers and mosses. At the same time, moose and caribou return from areas to the south and bears, squirrels, and other hibernating animals become active again. This season of relative warmth and productivity lasts about two months.

Some of the most unusual animals found in Alaska are the Dall sheep, which live in the Brooks Range among rocky slopes and open tundra. Found exclusively in Alaska and British Columbia, they are the only species of wild white sheep in the world. Alaska is also home to the migratory moose and caribou, which avoid the harsh winters of the North Slope by traveling south in the fall and returning in the spring. Two major herds of caribou, totaling approximately 370,000 animals, spend the summer in the North Slope. The porcupine caribou winters in Canada and migrates to the northeast area of the Slope. The Central Arctic caribou winter in Southeast Alaska, south of the Brooks Range, and migrate northwest into a summer spread of about 140,000 square miles in Northwest Alaska.

The unspoiled nature of the North Slope and surrounding areas allows the animals to follow the same routes every year, and it is difficult to predict how any disturbance or barriers in that environment, such as building crews or the pipeline itself, affect migration. Caribou, for instance, are easily deterred by anything in their way; faced with a fence or high brush, they will turn and follow the impediment to its conclusion rather than attempt to jump over or force their way through.

Fish also migrate to the North Slope area, although they reverse the patterns of the mammals and spend the winter in Alaska's rivers. Salmon, for instance, winter in pools of the Sagavanirktok River and migrate into the Arctic Ocean with the onset of spring. From mid-July to mid-August, the salmon travel back up the river, where they spawn between September and October. Their migration cycles and spawning beds could easily be disturbed by sections of pipeline or roads that cross rivers and streams, or by construction activity such as diverting streams to lay pipe underneath.

SOCIAL BOUNDARIES: THE TREATMENT OF NATIVE
ALASKANS AND THE USE OF NATURAL RESOURCES

When the United States bought the Alaskan Territory from Russia in 1867, no one dreamed of the resources that would be discovered there. Once the gold rush of 1899 brought the first major flux of outsiders to Alaska, though, the territory came to possess a mythic potential for non-Alaskans. Most have seen it as a place to make their fortunes, through gold, trapping, hunting, fishing, or, more recently, oil drilling. Others have valued Alaska for its vast tracts of unspoiled wilderness.

In their zeal to uncover Alaska's resources, however, newcomers have consistently failed to consider the rights of the Native Alaskans who have lived in Alaska for centuries. The native people who inhabited Alaska when Russian explorers first came to its shores in the 1700s included the Eskimos, who lived on the north and west coasts, and the Aleuts, who lived on the Aleutian Islands and the Alaskan Peninsula. Both the Eskimos and the Aleuts are believed to be descended from a group of people who lived on the land bridge that once connected Alaska and Siberia, and who migrated from the land bridge east to what is now Alaska and west to north-east Siberia. The native population also included the Tlingit, Haida, and Tsimshian Indians along the southeast coast, and the Athabaskan Indians in the interior.

At the time of the U.S. purchase of Alaska, most Native Alaskans lived the traditional lifestyles of their ancestors, hunting and fishing for a living and governing themselves through ancient tribal systems. Most white settlers and traders had little or no regard for the Natives' traditions or territories because they considered themselves racially and culturally superior. They took whatever they wanted or needed from the Natives, providing little or nothing in return. Some Natives responded by adopting the unfamiliar lifestyles and occupations of the outsiders, while others tried to withdraw from white civilization as much as possible in order to preserve their traditional ways.

In 1971, the U.S. government took steps to protect the rights of Native Alaskans. President Richard Nixon signed the Alaska Native Claims Settlement Act, which granted the Natives 44 million acres, or approximately 10 percent of the state. The act also reserved another 100 million acres for the Natives as federal protectorate land and granted them $962 million in cash. The act assigned the 44 million acres to twelve Native corporations, which were free to handle the land and associated resources as they saw fit.

Around the same time that Native Alaskans were gaining some control over their homeland, a group of oil companies was planning a pipeline to transport North Slope oil south to the Gulf of Alaska. Although its plans were stalled by a reluctant Congress, world politics soon altered the scene in the oil companies' favor. On October 6, 1973, Egypt and Syria invaded Israel. When the United States continued to support Israel, Saudi Arabia and the other countries in the Organization of Petroleum Exporting Countries (OPEC) protested by instituting a total embargo on oil exports to the United States. Because the United States relied heavily on OPEC countries for oil, the embargo precipitated monumental rises in oil and gasoline prices in the United States. Under pressure to aid the development of domestic oil reserves, Congress quickly passed the Trans-Alaska Pipeline Authorizing (TAPA) Act, authorizing construction of the pipeline, on November 16, 1973.

LOOKING BACK: THE HISTORY OF THE ALASKAN PIPELINE

The history of the Alaska pipeline reveals its intimate connection with the Alaskan environment. Environmental considerations shaped the way the pipeline was constructed, and, once built, the pipeline and its related oil fields and tankers changed the Alaskan environment in ways that are still being discovered.

Planning and Constructing the Pipeline

Once the oil companies realized how much oil was available in the North Slope, they began making plans to transport it and formed the Trans Alaska Pipeline System (TAPS) in February, 1969. One of TAPS's first decisions was to extend the pipeline from the oil fields of Prudhoe Bay to the city of Valdez, a port on the Gulf of Alaska where oil could be loaded onto tankers and transported by sea. The chief alternative to this plan had been a pipeline four times as long, extending from the Arctic through Canada's MacKenzie Valley to the U.S. Midwest. Although conservation groups preferred this longer pipeline because it would have eliminated the need to transport any oil by tanker and because it would have avoided the three geologically active fault zones crossed by the shorter pipeline, uncertainties about how Canadian laws and attitudes would affect such a plan led TAPS to choose the Prudhoe Bay-Valdez pipeline.

Soon after this decision, TAPS realized that its pipeline would be one of the first projects to have to conform to the National Environmental Policy Act (NEPA) of 1969, which went into effect on January 1, 1970. Among other things, NEPA required that an environmental impact statement (EIS) be prepared for any project using federal funds or undertaken on federal grounds. The EIS is supposed to reflect a thorough investigation of the project's potential impacts and is used to decide whether or not the project can be approved.

Besides having to comply with the new environmental regulations, TAPS faced a variety of lawsuits throughout the pipeline's development. The suits were usually filed by groups who were opposed to the pipeline or who objected to some aspect of its construction. For example, in early 1970, seven Native Alaskan villages sued TAPS, withdrawing the waivers they had signed in order to allow the pipeline to be built on their territory. They felt that Natives were no longer assured jobs associated with the pipeline as they had been promised when they signed the waivers. Three environmental groups, The Wilderness Society, Friends of the Earth, and the Environmental Defense Fund, also filed suit to reduce TAPS's right-of-way to one of 25 feet on either side of the pipeline, in accordance with a 50-year-old state law.

While TAPS struggled with political opposition and legal complications, its engineers faced perhaps the biggest challenge of all: how to build a pipeline that could transport oil at reasonably warm temperatures (since cold oil flows very slowly) without transferring heat into the surrounding environment and melting the permafrost. No one wanted to repeat earlier mistakes. In 1969, seeing the proposed pipeline as an opportunity to enhance Alaska's economic growth, the state's governor ordered that a construction highway be built along the proposed pipeline path. Failing to consider the delicate balance of the permafrost below the road's surface, the road crews simply cut down trees and stripped moss and grass from the ground along the roadway. Deprived of the plant life that insulated it, the permafrost thawed once spring arrived and the highway turned to mud. This incident served as a warning that the environment must be preserved not only for its own sake, but also to ensure the successful construction of the pipeline.

Originally, TAPS had hoped to bury most of the pipeline rather than to construct aboveground supports for the pipe. However, they found that adequately refrigerating underground pipe would be too costly and complicated except in areas where the permafrost was especially stable owing to constant cold temperatures. Eventually, they decided to bury only about 372 miles of the 800 miles of pipeline. They refrigerated the pipe, insulated it with polystyrene, and surrounded it with cooling pipes that circulated refrigerated brine.

The 423 miles of aboveground pipeline were laid on vertical support members, which consisted of teflon coated crossbeams held up by 18-inch diameter uprights drilled deep into the ground. The uprights were made of wood because it conducts heat very poorly. Many uprights also contained sealed ammonia tubes to absorb any heat emitted from thawing permafrost and transmit it into the air. In order to allow for the pipe constricting and stretching with changing temperatures, each section was built at a slight angle to its adjacent sections.

In response to environmental regulations and concerns about the pipeline's effects on surrounding plants and wildlife, the oil companies commissioned a variety of ecological studies on the areas to be affected by the pipeline. These studies resulted in adaptations such as constructing over 800 crossings for migratory animals into the aboveground pipeline. When temporarily diverting streams in order to lay pipes underneath, construction crews were also careful to work at places and times that would not disturb spawning grounds of salmon or other fish.

After TAPS secured the approval of the Department of Transportation and the Alaska Pipeline Office, the pipeline was ready for use on June 20, 1977. Putting the pipeline into use was not without incident, however. As oil began to flow, an explosion and fire took place at a pump station near Fairbanks, killing a pipeline technician and putting the station out of service for seven months. In addition, several leaks occurred during the pipeline's first few years; one was attributed to sabotage and the others were traced to the settlement of the pipeline itself.

Assessing the Pipeline's Environmental Impact

According to most studies conducted during the 1980s, the precautions taken by TAPS to preserve the environment surrounding the pipeline were successful. Almost all migratory animals seemed to have adapted to crossing the pipeline, and the fish seemed to be maintaining their natural cycles of behavior despite the disturbance to the rivers and streams where they spawned. The "corridor" through which the pipeline passes appeared revegetated, and the permafrost along the pipeline seemed intact.

Although Alyeska, the company that managed the Pipeline, and the oil companies boast that the pipeline has had no negative effects on the Alaskan environment, some environmentalists believe that it is simply too soon to pass judgment and that we must wait to assess the long-range effects of the pipeline. Although the corridor has been largely revegetated, the construction of the pipeline and the development of the oil fields have disrupted thousands of acres of wilderness habitats. The human activity in the Prudhoe Bay area has resulted in some obvious impacts as well. Most of the buildings, oil wells, and roads sit on gravel pads that help prevent the permafrost from melting, but that also restrict the natural flow of water and cause excessive flooding in some arctic wetlands. Human activity in the Arctic also causes and accelerates thermokarst, a localized thawing of ground ice which causes a depression in the ground. These depressions often fill with water, which increases the ground's heat absorption, leading to further thawing.

Although the degree to which the oil fields and finished pipeline have disturbed Alaska's environment is still a matter of debate, the 1989 oil spill in Prince William Sound dramatically illustrated the potential environmental costs of our quest for oil. On March 24, the *Exxon Valdez* ran aground on Bligh's Reef, spilling more than 11 million gallons of crude oil into the pristine Alaskan environment. Eventually, although 2.6 million gallons of that oil was recovered by Exxon, an untold amount dispersed throughout the ocean and washed ashore.

In the face of unprecedented public outcry, Exxon funded a massive campaign to attempt to repair the damage: they set up centers to rescue oil-drenched birds and otters, paid workers to clean over 1000 miles of beach and rocks, and reimbursed fishers for their lost catches. Despite these efforts, approximately 1000 otters and 33,000 birds were known to have died, and countless more simply sank to the bottom of the ocean. Although the cleaned beaches appeared free of oil on the surface, a thick layer of oil-soaked sand and gravel lay hardening underneath. Only time will tell what permanent effect the spill has had on water, soil, and the plants and animals that depend upon them.

Even worse, this highly publicized spill was not an isolated incident. Subsequent spills included significant incidents near New York, New Jersey, and California. In June 1989, the US EPA issued a report that revealed the oil industry's hundreds of violations of state and federal regulations as well as the serious environmental damage that had been caused by oil development and transportation. In the aftermath of these sobering incidents, the nation began to reevaluate its methods of energy use and development. An integral part of that ongoing reevaluation is closer scrutiny of the Alaska pipeline.

The highly publicized oil spill prompted Alyeska to fund a permanent citizens' watchdog group at an annual cost of $2 million. Established in 1990 to monitor Alyeska's shipping operations, the fifteen member Regional Citizens Advisory Committee (RCAC) is made up of representatives from Alaskan towns, fishing organizations, and Native groups. Alyeska has no power to cut off the group's funding or to choose members. If Alyeska disagrees with any of RCAC's recommendations, it must provide a written defense of its refusal. Any disagreements are settled by arbitration, and the results, because they are binding, avoid lengthy, expensive lawsuits. One of RCAC's first tasks was to review an oil spill response plan submitted by Alyeska to the State of Alaska.

Although RCAC was a positive step toward protecting Alaska's environment, another major concern came into the public eye early in 1990: the detection of external corrosion along the pipeline. The public became concerned that corrosion could eventually lead to a blowout, polluting the surrounding wilderness and forcing an expensive shut-down of the line. Alyeska maintains that a blowout precipitated by corrosion is impossible because detection equipment is so advanced that problems are quickly identified and corrected. In 1991, Alyeska replaced 8.5 miles of underground pipe at Atigun Pass; the pipe was located in a floodplain and showed signs of external corrosion. In addition, Alyeska plans to spend $800 million through 1993 to repair corrosion along the pipeline, at pump stations, and at the Marine Terminal in Valdez.

Critics say that most corrosion could have been avoided if Alyeska had taken proper precautions during construction. They charge that Alyeska rushed construction and never conducted planned tests of corrosion protection because the oil companies lost money every day the completion of the pipeline was delayed.

In response to the *Exxon Valdez* incident and the discovery of corrosion, the United States General Accounting Office conducted a study of how well the five principal state and federal agencies responsible for regulating the pipeline system have performed their duties of ensuring the system's operational safety, oil spill response capabilities, and ability to protect the environment. In a July 1991 report, the GAO determined that these agencies had not adequately overseen TAPS; instead, they relied on Alyeska to police itself. However, in January 1990, the regulators had established a joint office to provide for more effective TAPS oversight. The GAO believes that central leadership and a secured funding source may help ensure that this office provides adequate oversight.

LOOKING AHEAD: THE FUTURE OF THE
PIPELINE AND THE ALASKAN ENVIRONMENT

Although the pipeline itself seems to have left most of Alaska's natural environment relatively unchanged and the memory of the *Valdez* disaster is slowly fading, oil development still poses two types of threats to the Alaskan environment. The first, of course, is more oil spills or an oil blowout. The second is the attempt by oil companies to obtain drilling rights in areas protected by the government as wilderness reserves.

Currently, the pipeline transports almost 2 million barrels of oil a day, roughly 20 to 25 percent of annual U.S. oil production, but pipeline officials estimate that production will decline to 500,000 barrels a day by the year 2000 and dwindle even further after that. To offset dwindling supplies, oil companies are clamoring to extend their drilling rights into the the Arctic National Wildlife Refuge (ANWR), a 19 million acre wildlife refuge just east of the developed Prudhoe Bay area. ANWR was expanded to its present size by the passage of the Alaska National Interest Lands Conservation Act in 1980. A victory for conservationists, this act created numerous parks, wilderness areas, and wildlife refuges throughout the state. However, supporters of the bill had to make certain compromises to ensure its passage. Among these was the failure to gain wilderness designation for ANWR even though it is the last fully intact arctic ecosystem in the United States. Without such designation, ANWR was left open to drilling for oil and gas with the permission of Congress.

Although the *Valdez* oil spill temporarily put a damper on the demands to drill in ANWR, the oil industry seized on the 1991 Persian Gulf War as an opportunity to argue for the necessity of exploiting the area's as yet unproven oil resources as an alternative to reliance on foreign oil. Despite the oil companies' claims that ANWR and other protected areas could hold an undiscovered wealth of oil, most studies predict that ANWR probably contains only about 3.2 billion barrels of oil, enough to meet U.S. oil demand for about six months. In addition, to make those 3.2 billion barrels economically worth recovering, oil prices would have to be roughly twice their current level.

Moreover, some environmentalists argue that any additional drilling in the North Slope will simply contribute to overall oil consumption and thus add to related problems such as pollution and the greenhouse effect. Others worry that, although the Prudhoe Bay oil activities haven't resulted in major consequences for the Alaskan environment to date, adding the proposed oil fields to the existing ones could have a serious cumulative effect.

In addition, many scientists point out that the area within ANWR seems to be more vulnerable to environmental damage than the Prudhoe Bay area. The refuge has a summer population of caribou fifteen times greater than that of the Prudhoe Bay area. Other mammals integral to the ANWR ecosystem are beaver, wolves, musk oxen, brown bears and polar bears. Over 150 types of birds, including geese and the threatened peregrine falcon, nest in ANWR during the summer, and its waters are home to great numbers of fish. Exploring and constructing oil fields could seriously disrupt the life cycles of these populations, and drilling off the shores of the refuge could disturb populations of threatened bowhead and gray whales. Any impact on caribou populations could harm the Gwich'in Indians of the Athapascan tribe which depend on the caribou for food. Offshore drilling could affect the Inupiats of the North Slope, who rely on whale and seal populations.

Even the Department of the Interior, frequently in agreement with the oil industry, has prepared an Environmental Impact Statement (EIS) that predicts that drilling would have major negative effects on the caribou and musk oxen of ANWR and would moderately interfere with populations of wolves, brown bears, and snow geese. In addition, the EIS warns, drilling could ruin the subsistence lifestyle of the Inupiat people and would bring with it increased noise, loss of wilderness, and the possibility of hundreds of minor fuel spills.

Although the oil industry presents the ANWR issue as a choice between the violation of this wildlife refuge and an ever increasing reliance on foreign oil, there are other viable options. Simple conservation measures could easily save more oil than we could possibly find in ANWR. By raising automobile mileage standards to 40 miles per gallon, we could save more than 2 million barrels of oil a day, not much less than the estimate of 3.2 million barrels total for the oil to be found in ANWR. In addition, increased government funding for energy alternatives to nonrenewable sources would help to provide us with *lasting* and more environmentally benign forms of energy. The environmental record of the Alaskan oil industry, combined with the threat to an area that is an irreplaceable and unique ecosystem, should make it clear that we cannot continue our present patterns of energy consumption indefinitely.

SUGGESTED READING

Adler, Jerry. "Alaska After." *Newsweek*. September 18 1989. 50-62.

Air Pollution: Status of Dispute over Alaska Oil Pipeline Air Quality Controls. Washington, DC: U.S. General Accounting Office, 1988.

Allen, Lawrence J. *The Trans-Alaska Pipeline*. Seattle: Scribe Publishing, 1975.

Baum, Dan. "Trespassing the Silent World." *Buzzworm: The Environmental Journal*. July/August 1991. 30-37.

Berry, Mary Clay. *The Alaska Pipeline: The Politics of Oil and Native Land Claims*. Bloomington, IN: Indiana University Press, 1975.

Borrelli, Peter. "Troubled Waters: Alaska's Rude Awakening to the Price of Oil Development." *The Amicus Journal*. Summer 1989. 12-20.

Dixon, M. *What Happened to Fairbanks? The Effects of the Trans-Alaska Oil Pipeline on the Community of Fairbanks, Alaska*. Boulder, CO: Westview Press, 1978.

Freeman, David. "Saving the Alaska Pipeline." *Popular Mechanics*. August 1990. 68.

Hedin, Robert, and Gary Holthaus, eds. *Alaska: Reflections on Land and Spirit*. Tucson, AZ: University of Arizona Press, 1989.

Hodgson, Bryan. "Alaska's Big Spill: Can the Wilderness Heal?" *National Geographic*. January 1990. 5-43.

Hoshino, Michio. "Caribou—Majestic Wanderers." *National Geographic*. December 1988. 846-857.

Mead, Robert Douglas. *Journeys Down the Line: Building the Trans-Alaska Pipeline*. Garden City, NY: Doubleday and Co., 1978.

Rauber, Paul. "Last Refuge." *Sierra*. January/February 1992. 35-43.

——. "The Man Who Knew Too Much: Alyeska Plugs a Leaky Pipeline with its Own Private FBI." *Sierra*. March/April 1992. 76-77.

Reynolds, Brad. "Athapaskans Along the Yukon" *National Geographic*. February 1990. 44-73.

Roscow, James P. *800 Miles to Valdez: The Building of the Alaska Pipeline*. Englewood Cliffs, NJ: Prentice-Hall, 1977.

Seligman, Dan. "The Accident That Waited to Happen." *The Amicus Journal*. Summer 1989. 24-29.

Trans-Alaska Pipeline: Regulators Have Not Ensured That Government Requirements Are Being Met. Washington, DC: General Accounting Office, 1991.

Vincent, Brian. "Reflections on the Exxon Valdez: Will the Real Drunken Sailor Please Stand Up?" *E Magazine*. May/June 1990. 58-59.

Walker, D.A., et al. "Cumulative Impacts of Oil Fields on Northern Alaskan Landscapes." *Science*, November 6, 1987. 757-760.

Yergin, Daniel. *The Prize: The Epic Quest for Money, Oil, and Power*. Simon and Schuster, 1991.

CHAPTER 9

CONVENTIONAL FOSSIL FUELS: STRIP MINING FOR COAL IN APPALACHIA

In the United States, most of us take our easy access to relatively inexpensive sources of energy for granted. We expect the lights to turn on when we flip the switch, and we would be shocked if our local gas station closed because of lack of gasoline. However, roughly 90 percent of this readily available energy supply comes from sources that are dwindling and nonrenewable: fossil fuels, principally coal, petroleum, and natural gas.

As the term "fossil" suggests, these fuels are the fossilized remains of organic matter. When a fossil fuel is burned, or combusted, the chemical bonds that bind its molecules together are broken and energy is released. This energy, whether it is contained in a tankful of gas or a lump of coal, was once sunlight, captured by green plants through photosynthesis. When dead plants and the animals that ate them settled to the bottoms of seas, swamps, and other wetlands millions of years ago, they formed thick layers of organic matter rich in carbon. A process of decomposition, coupled with heat from the Earth's interior and pressure from water and accumulating sediments, slowly transformed the organic material into fossil fuels.

The processes that formed fossil fuels continue today, as decaying plant materials accumulate in coal-forming environments like South Florida's Everglades. But because those processes occur on a geologic time scale, the world's present deposits of coal, petroleum, and natural gas are all we can ever expect to use. Because fossil fuels are essentially nonrenewable and finite in supply, a crucial issue facing us today is how to ease and eventually eliminate our reliance on these fuels and how to develop renewable, environmentally sound energy sources in their place. In the meantime, we must minimize the harm to human health and the environment which can result from the recovery and use of fossil fuels.

Because the environmental conditions that led to the preservation and fossilization of organic matter did not occur everywhere, fossil fuels are not distributed evenly beneath the earth's surface; here in the United States, we have a fairly limited domestic supply of oil. However, we do have significant deposits of coal. Mining this coal has involved a host of problems, including ravaged landscapes and the exploitation of the people who worked in the mines. In addition, the combustion of coal often produces sulfur, one of the factors that produce acid precipitation, and always produces carbon dioxide, the principal cause of global warming.

There are three major coal producing areas in the United States: the western coal fields, the midwest basin, and the mountains of Appalachia. Although all these areas share problems related to coal mining, the unique physical, biological, and social characteristics of the Appalachian region make that area an especially striking example of the needs, costs, and dangers surrounding our reliance on the most abundant U.S. fossil fuel. The Appalachian Mountain range extends from New York to Alabama, and its major coal producing states are Pennsylvania, Ohio, West Virginia, Virginia, Kentucky, and Tennessee. Roughly half of Appalachia's coal production is obtained by strip-mining, or removing all the overlying earth in order to expose the seam of coal. Although strip mining is practiced throughout Appalachia, most of it occurs in southeastern Ohio, eastern Kentucky, West Virginia, and Alabama.

Strip mining has caused extensive damage to the environment, leaving water polluted and land literally stripped of the layer it needs to sustain life. In addition, the mining industry has too often acted without thought for the people who live in the area being mined. Although major legislation passed in the 1970s has required strip mining companies to eliminate many of the practices that have caused damage to the environment and those who depend on it, our society still needs to deal with vast areas of unreclaimed land as well as dangerous violations of regulations that threaten to leave even more ruined land.

PHYSICAL BOUNDARIES: THE MOUNTAINS AND COAL OF APPALACHIA

The geographical area referred to as the Appalachian Mountains is a combination of three distinct geological formations. The main mountain range, known as the Older Appalachians, is actually a chain of mountain groups extending from Maine to Georgia. These groups include the Green Mountains in Vermont and the Unaka Mountains in Tennessee. Thought to be the oldest range in North America, these mountains have eroded substantially since a geological upheaval thrust them from the sea about 500 million years ago.

To the west of this chain, a newer range known as the Allegheny Mountains extends from southern New York to Alabama; this range includes the Cumberland Mountains of Kentucky. In addition, many small ranges occur at right angles to the Older Appalachians and the Alleghenies, especially in the southern region of Appalachia. These cross ranges include the Black Mountains of North Carolina, where the Appalachians reach their highest peak: Mt. Mitchell, which has an elevation of 6,684 feet. The Great Appalachian Valley, between the Alleghenies and the Older Appalachians, extends from the St. Lawrence Valley in Canada to Alabama.

The Appalachian region is drained by a number of major rivers, including the Delaware, the Susquehanna, the Potomac, the Kanawha, the Cumberland, and the Tennessee. In addition, numerous smaller streams are found throughout the region. The stream valleys and floodplains of the region are especially susceptible to flooding, and the mountains' steep slopes are subject to frequent landslides. Unfortunately, mining activities often encourage erosion and other conditions that increase the likelihood of landslides even further.

The Appalachian region provides approximately 70 percent of the bituminous coal produced in the United States. Bituminous, or soft, coal is the most common variety, accounting for over half of the U.S. reserves. Although its heating value is lower than that of anthracite, or hard coal, bituminous coal is used almost exclusively for electric power because anthracite coal is extremely rare (it comprises only 1 percent of our reserves) and expensive. The two other major varieties of coal are subbituminous and lignite coals, which account for 48 percent of our reserves. They have one advantage over bituminous coal; although over 50 percent of bituminous coal reserves contain high levels of sulfur, which contribute to acid precipitation when the coal is burned, subbituminous and lignite varieties contain very little sulfur. Unfortunately, their heating values are considerably lower than that of bituminous coal.

Although the coal-rich areas in the midwest and west are generally flatter and easier to mine than the steep Appalachian mountain ranges, Appalachia's coal remains economically competitive because much of it is high in energy while containing low amounts of ash and less than 2 percent sulfur (relatively low for bituminous coal). A few areas do produce higher sulfur coal; Ohio's coal is worst in this respect, usually containing more than 3 percent sulfur. However, mining companies still find buyers for high-sulfur coal by mixing it with a lower-sulfur variety or by selling it to electrical plants that have scrubbers to remove toxic emissions. Unfortunately, until the full implementation of the 1990 Clean Air Act, some plants will still be able to burn high-sulfur coal without using scrubbers, resulting in emissions that cause acid rain.

Another characteristic affecting Appalachian coal production is the terrain, which determines the type of strip mining that may be practiced there. Area mining, which removes overburden (the soil, plants, and everything else above the coal seam) in box like cuts and then deposits it back in the previous cut, is suitable for flat terrain such as the midwest basin and western coal fields. Area mining can be adapted for some mountaintops, too; the mountaintop is removed, producing a level surface on which to work. But in Appalachia, where most areas have slopes of over 12°, the chief method of surface mining is contour mining. Contour mining operations cut into the hillside, leaving a shelf or bench in the slope. These cuts follow the coal seam, producing a long, snake-like gouge that winds across the terrain. Conventional contour mining creates a heaped up spoil bank of removed overburden in front of the shelf and a highwall at the back.

BIOLOGICAL BOUNDARIES: ECOSYSTEMS AFFECTED BY STRIP MINING FOR COAL

Much of Appalachia is forest or farmland, providing habitats for numerous plant and animal species. Large mammals include deer, black bears, bobcats, and foxes. Although the large size of the region and the variety of altitudes found there result in a wide range of vegetation, dominant tree species include spruce, white pine, cedar, hemlock, white birch, maple, and basswood. Ash and hickory are prevalent in the southern area of Appalachia. Barley, potatoes, wheat, alfalfa, and apples are major crops grown in the northern region, whereas corn and tobacco are prominent in the south.

Strip mining can endanger these communities of plants and animals in many ways. First of all, mining operations destroy the vegetation in the area, ruining it as a habitat for the animals. The disrupted land may act as a barrier for migrating animals. After the seam is mined out, water and soil pollution arising from strip mining can render an area unsuitable for fish, other wildlife, and people.

SOCIAL BOUNDARIES: THE HUMAN
COSTS OF STRIP MINING IN APPALACHIA

Appalachia is not only a distinct geographic region, but it is culturally unique as well. Stereotypes of the people who inhabit Appalachia abound. These stereotypes range from the negative picture of large families preoccupied with feuding and making moonshine to romanticized images of folk of descendants of Scottish and British immigrants perpetuating Elizabethan culture in the midst of modern America. Most experts on Appalachia would respond to these stereotypes by first pointing out that there is no one type of Appalachian inhabitant. Not all Appalachians live in the mountains; the region includes flat, accessible land and thriving urban centers. These more urban Appalachians tend to have higher incomes, lower unemployment, more comfortable living conditions, and higher educational levels than those in the mountainous region.

Although many of the mountain-dwelling Appalachians do come from Anglo-Saxon backgrounds and tend to be strongly religious, they do not conform to any one description. Many of the characteristics they do share can be at least partially attributed to environmental factors. Historically, the mountainous topography has tended to make communities small and inaccessible, fostering a sense of independence and a strong family loyalty that exist to this day. The area's steep slopes and narrow valleys are flood-prone, and hard to farm; coupled with a short growing season, these factors help to keep residents from becoming especially prosperous.

One thing the people of Appalachia do have in common is the overwhelmingly negative heritage of the coal mining industry which has exploited both them and their land. Before World War II, coal met over 50 percent of the country's energy needs. At that time, most coal in Appalachia was mined underground. Underground mines posed various dangers to miners: asphyxiation by toxic coal fumes, mine collapse, and black lung, caused by inhaling large amounts of coal dust. After the war, the advanced earth-moving technology developed for military purposes was adapted to commercial enterprises, and strip mining became commonplace. Although it posed a much greater threat to the environment than underground mining, strip mining was initially relatively free from environmental regulations.

Strip mining allowed companies to extract coal more efficiently—80-90 percent of a coal seam rather than the 50-75 percent available in an underground mine. Suddenly, seams which were too weak or thin to be mined underground became economically viable. Working above ground minimized health and safety threats, but the shift to machine-intensive strip mining left tens of thousands of miners jobless. Between 1949 and 1985, domestic coal production doubled, but employment dropped from almost 400,000 miners to less than 200,000. An exodus from Appalachia took place, as many of the unemployed left the region and headed north to work in the factories in cities like Detroit, Michigan, and Cincinnati and Hamilton, Ohio. Unemployment separated families and friends, leaving once-proud mountain people isolated in unfamiliar cities, and the cultural and social fabric of Appalachian life began to tear.

Between World War II and the early 1970s, coal's share of the domestic energy market dropped as the use of oil and natural gas increased. The 1973 OPEC oil embargo revived the coal industry, at least temporarily. By the end of the decade, coal from some mines in Appalachia was selling for as much as $100/ton. As a result of the increased demand, mining in the Rocky Mountain region expanded considerably, and Wyoming soon overtook Kentucky and West Virginia as the nation's leading coal-producing state. Even so, by 1975, approximately half of the coal (640 million tons) mined in the United States was obtained by strip mining, and Appalachia provided about 50 percent of that total.

The "boom market" of the 1970s collapsed in the early 1980s. The very high prices of the 1970s had spurred foreign buyers such as Japan and Taiwan to turn to other coal exporting nations, particularly Australia and Canada. United States coal exports, which had reached an all-time high in 1981, dropped 30 percent by 1984. Environmental concerns caused increasing numbers of utility companies to switch to cleaner burning oil and natural gas as prices for these commodities dropped. Perhaps the most devastating blow for Appalachia was the decline in the nation's steel industry. Steel is produced using what is known as "met coal" (metallurgical coal), most of which is obtained from Appalachian coal fields. In 1974, consumption of met coal was 90 million tons/year; by 1985, it fell to 40 million tons. By 1995, consumption is expected to be at just 32 million tons/year.

Coal mining's legacy, which began generations ago but was forged during the period since World War II, can be seen in Appalachia today. Thousands of acres of land strip-mined before the 1970s remain ruined, with once productive, balanced environments reduced to eroded, scarred wastelands. According to the *1991-1992 Green Index*, a state-by-state guide to the country's environmental health, West Virginia ranks 50th on the sulfur dioxide indicator, generating twice as much of this precursor of acid precipitation as does the next worst state, Indiana. Not surprisingly, a fifth of the lakes and streams in Appalachia, as far south as Tennessee and North Carolina, have acidic pH levels. Kentucky and West Virginia also rank among the worst ten states for carbon emissions, occupational deaths, and high-risk jobs. Ohio and Pennsylvania also rank poorly in terms of sulfur and carbon emissions. Water pollution is another serious problem in Appalachia. West Virginia, for example, ranks last among all states in terms of surface water quality, forty-eighth in terms of overall water pollution, forty-third in spending for water quality, and forty-sixth in the investment needed for adequate sewage facilities.

What do all these numbers mean for the people of Appalachia? Although strong unions have enabled working miners to secure relatively good wages and health benefits from employers, this is little comfort in a region plagued by poor schools, severe pollution, weak enforcement of safety regulations, poverty, high unemployment, and a stagnant economy. Many families in West Virginia have no running water, nor do they have the money to install pipes or septic tanks. Some families still haul water from nearby creeks, which may be contaminated with mining, industrial, or household waste, or with acid precipitation. Kentucky, West Virginia, and Alabama rank forty-sixth, forty-seventh, and forty-eighth, respectively, in terms of community health. A second exodus is taking place in Appalachia, but many of those now leaving the region are heading south, for the manufacturing plants of North Carolina and Georgia.

LOOKING BACK: THE HISTORY OF STRIP MINING IN APPALACHIA

The history of strip mining in Appalachia is largely one of greed and environmental destruction. However, within the past few decades, growing environmental consciousness in the United States has catalyzed a shift toward more responsible mining practices.

Exploiting the Land

Water pollution has been one of the most serious and far-reaching problems caused by conventional contour mining. Coal and the soil around the coal seam contain materials that produce acids (usually sulfuric acid) when exposed to oxygen, and the coal and acidified soil may contaminate nearby waters.

The practices of conventional contour mining made it easy for water to erode the earth in and around the pits. Once removed from a cut, overburden was piled into increasingly larger spoil banks that overflowed down the hillside. These piles encouraged erosion and landslides, which carried acid, silt, and heavy metals into nearby water sources. The highwall at the back of the cut was also eroded by water running from the top of the hill or mountainside. When silt from the eroding highwall clogged drainage ditches, water settled in the bottom of the pit. All of this contaminated water eventually entered groundwater, streams, and lakes.

Once it reached streams and lakes, sedimentation, acid, and other pollutants caused widespread damage. They contributed to the erosion of bridges and roads, clogged culverts, and prevented dams from controlling floods and storing water. Toxic substances contaminated wells, springs, and other sources of drinking water. As of 1982, over 11,000 miles of U.S. waterways had been polluted from acid run-off and siltation caused by both strip and underground mining.

Contour mining also seriously damaged the ability of the land to support life. Even in rare cases where mining operations deposited spoil in previously mined cuts, the topsoil was usually completely destroyed. Because it was the first layer of overburden removed, the good topsoil was invariably ruined by the acid-contaminated earth piled on top of it. As a result, strip-mined land could almost never regenerate any plant life and was covered either with empty pits or with ridges resulting from dumping piles of spoil into cuts. Mining roads, ditches, and leftover highwalls often disturbed the original drainage patterns of the area.

The land disruption and water pollution resulting from conventional contour mining adversely affected wildlife. Although most wildlife fled an area being mined, many burrowing animals were killed once mining operations began. Because the mining process destroyed their natural habitats, animals could almost never repopulate a strip-mined area. Strip-mined areas could also disorient migratory animals from their traditional patterns, and extensive mining operations substantially reduced feeding areas.

Aquatic organisms were perhaps most affected by mining operations. Eroding soil from spoil banks washed into nearby streams and rivers and accumulated as sediment on their bottoms. Sediments in streams have been known to accumulate to a depth of 18 inches or more in the first six months of a strip mining operation. Such build-up reduced the water carrying capacity of affected streams by as much as two-thirds. The first creatures affected by accumulating sediment were those who lived on the stream's bottom, such as clams, snails, crustaceans, and some insects. Populations of fish that fed on these bottom-dwellers or that laid their eggs on stream bottoms were soon endangered as well. Although some species that fed and laid eggs close to the water's surface survived the effects of sedimentation and even flourished in the absence of competitor species, the resulting stream ecosystems were dramatically imbalanced. Unfortunately, the negative effects of sedimentation do not cease once mining ends. A study of streams affected by strip mining in east-central Kentucky showed that even seventeen years after mining in the area ceased, fish populations had made only minor advances towards recovery.

Although accumulating sediment means long-term disaster for stream ecosystems, acid and other pollutants that can wash out of strip mines can turn streams and rivers into death traps much more quickly, sometimes exterminating entire populations. For example, in August 1966, heavy rains washed an especially large amount of coal-associated acid into the Allegheny River near Sharpsburg, Pennsylvania. The resulting toxicity killed one million fish. Studies show that even small amounts of less lethal pollutants, like heavy metals, can alter the patterns of migratory fish; they simply turn around rather than moving through the polluted area to reach their spawning grounds.

Conventional contour mining was also disruptive for the human population of the area. Mining operations brought noisy machinery and constant traffic. Sometime whole communities had to move because of poisoned water sources and ruined farmland. Even in less extreme cases, mining companies left communities surrounded by a scarred wasteland. Abandoned mining areas also often included dangerously toxic piles of abandoned coal refuse. These extremely flammable piles sometimes caught on fire, emitting toxic fumes like sulfur dioxide into the atmosphere.

Although the process of unregulated mining was almost always devastating for nearby communities, especially ruthless was the practices of strip-mining land to which companies held "broad form" deeds. Such a deed gave its owner mineral rights to land owned by someone else. Most landowners signed such deeds before the advent of strip mining, thinking an underground mine would not disrupt their property. However, many of these deeds fell into the hands of strip mining companies that simply unearthed their coal and left the owners with property unfit for any productive use. Some states, like Kentucky in 1987, have passed laws prohibiting strip mining by broad form deed. But for residents who suffered such indignities as watching coal shovels unearth family graveyards to get at the coal beneath them, the legislation came too late.

Protecting the Land

The 1960s and early 1970s witnessed increasingly strident demands for federal protection against the exploitation of the land by mining companies. Although some states developed their own regulation programs, federal laws were proposed unsuccessfully until 1977, when President Carter signed the Surface Mining Control and Reclamation Act (SMCRA), a huge step forward for conscientious resource management.

Under the act, each state established its own regulatory program under federal guidelines. Although the resulting regulations varied from state to state, all programs protected certain areas, such as wilderness preserves, from strip mining and required companies to obtain permits to strip-mine an area. To do so, mining companies had to show that they would prevent damage to water sources and that they had feasible plans for reclamation, the restoration of the land to its pre-mining level of productivity. For each site, the company had to post a bond equal to the estimated cost of reclamation.

To comply with the new regulations, strip mining operations quickly came up with simple adaptations that made contour mining much less devastating to the environment. Diversion ditches, which ran along the hillside above the top of the highwall, channeled water around the mining area. The "slope reduction" method simply spread the overburden over a larger area than before, reducing the slope and size of the spoil bank to prevent landslides and excessive erosion. The haulback technique (similar to area mining) deposited overburden into previously mined cuts, also reducing spoil erosion by eliminating spoil banks and stabilizing the highwall.

Strip mining companies also had to develop reclamation techniques. Most states required that the land be restored to its pre-mining level of productivity, including a self-reproducing plant community. The first basic step in reclamation is grading by bulldozers, which fill in the pits that were not filled by haulback techniques or some other method of integrated reclamation. Bulldozing smooths the area to match the contours of the land. Many states, including Ohio, require back-to-contour reclamation. To successfully reclaim an area, mining companies also have to be careful to store the topsoil separately from the rest of the spoil; then they can deposit the uncontaminated topsoil over the land after it has been graded.

The final, and perhaps most difficult, step in reclamation is revegetation. Plants must be carefully chosen based on an area's soil type, chances of erosion, and proposed post-mining use. Then vegetation must be monitored to make sure a stable community has been established. Reclaimed land can be used for almost anything; currently, reclaimed strip mines have been used for agriculture, forestry, recreation, and wildlife habitats, as well as residential, commercial, and industrial building. The Appalachian region is the national leader in number and diversity of post-mining land uses.

The SMCRA also established the Abandoned Mine Reclamation Fund, designed to encourage reclamation of strip mines abandoned before SMCRA went into effect. To obtain money for the fund, the government taxes both underground and surface coal production. Then both state and federal governments award contracts to currently operating mining companies to reclaim old mine sites. For instance, Waterloo Coal, which operates in Jackson County, Ohio, has contracted to reclaim an abandoned area near their current site in Oak Hill, Ohio. A small, privately owned company, Waterloo is indicative of a growing trend toward socially and environmentally responsible mining. Waterloo maintains a good relationship with the surrounding communities. Many community members find work with Waterloo, from life-long employees who run mining equipment to high school and college students who spend their summers spreading topsoil, seeding, and helping with reclamation in various ways. In its current operations, Waterloo successfully prevents water pollution and facilitates reclamation by using improved methods such as the haulback technique.

LOOKING AHEAD: RESTORING AND PROTECTING THE APPALACHIAN REGION IN THE FUTURE

Because of the 1977 Surface Mining Control and Reclamation Act, strip mining today is much less of a threat to the environment than it was in the days when mining companies simply abandoned exhausted sites. However, current methods of pollution control and reclamation are still imperfect, and vast areas of Appalachia still remain unreclaimed. In Ohio, for example, 210,000 acres still required reclamation as of 1987, while an additional 6000-7000 acres remained covered with coal waste. In addition, some mining companies still resist complying with environmental regulations.

Although mining companies have successfully reclaimed land all over the world, other efforts have failed or have only partially helped to restore the original productivity of strip-mined land. Reclamation is especially difficult in the more mountainous areas of Appalachia because the steep slopes encourage erosion and toxic run-off. As a result, the soil suffers from severe nutrient deficiencies and high concentrations of toxins, and it becomes very complex and expensive to condition the soil to a point where it can sustain vegetation. The continuous erosion also prevents plant growth that would help retain soil.

Other areas have revegetation problems because of inadequate topsoil. Even though most current strip mining operations store topsoil separately to prevent its contamination by acid, other damage still occurs. Simply digging up the topsoil breaks down the natural organic structure of the soil and destroys many of the microorganisms necessary for healthy soil. In many cases, the earthmoving machinery that replaces the topsoil packs it down so solidly that neither plant roots nor water are able to penetrate it. In addition, the soil below the topsoil is almost always very acidic; companies must frequently apply lime and fertilizers so that plants can grow. Unfortunately, most companies abandon such efforts when their period of legal responsibility ends if they do not own the land. Left on their own, some revegetated plant communities die out.

In addition, many companies are not providing strip mined areas with the biological diversity they had before mining, finding it easiest to revegetate with only one species of grass, legume, or tree. Most areas need more variety to become self-sustaining ecosystems. This problem is especially evident in the widespread conversion of forested land to pasture and hayfields through mining and reclamation. Although pasture and hayfields provide habitat for substantially fewer species than forests, they are attractive to owners who can reap greater economic benefits from them. Other revegetation efforts fail because of inappropriate plant choices; for instance, easily eroded areas need plant communities that can take root and spread quickly.

Research will help find answers to problems with topsoil and vegetation. Fortunately, some scientists are experimenting with different species of plants to find breeds which are especially good at preventing erosion and growing in acidic soil. Compost and sewage sludge is effectively being used to replenish nutrients on reclaimed land. In the area of computer modeling, designers are developing software to help monitor and prevent stream pollution from acid run-off.

But research is only part of the answer. The old attitude by which the land is viewed as something to exploit for human gain is alive and well: many people still oppose the Surface Mining Control and Reclamation Act and other efforts to make strip-mining safer for the environment. Powerful lobbyists for coal companies and other interest groups claim that some regulations are too harsh and unnecessary. Between 1981 and 1983, Interior Secretary James Watt made many changes that environmentalists see as regressions, such as weakening air quality controls and lowering reclamation standards.

In addition, many major environmental groups, such as the National Wildlife Federation, insist that the government is not adequately enforcing SMCRA. In fact, as of 1987, approximately 4000 strip mining sites were operating without the precautions required by SMCRA. Two thousand four hundred operations had ignored warnings to halt illegal operations, and the Office of Surface mining had failed to collect between $150 and $180 million in fines from violators. A serious shortage of federal and state mine inspectors contributes to lack of enforcement. Critics claim that state enforcement is also ineffective because state governments are hesitant to offend the coal companies that provide them with so much tax revenue. Another problem insufficiently addressed by state and federal authorities is the uncounted numbers of fly-by-night "wildcat" operations that have no regard for environmental regulations at all.

Other mining companies that don't actually violate the law find legal ways to get around it. Until 1987, for example, because SMCRA exempted small operators (under two acres) from some environmental restrictions, some large companies divided themselves into many two-acre operations to legally bypass regulations. Although a 1988 repeal of the two-acre exemption put an end to this abuse, others remain. In some cases, mining operations avoid reclaiming areas by forfeiting bonds which are substantially less than the cost of cleaning up. For example, Skyline Coal Co. in Kentucky put up a $6900 bond for a site they failed to reclaim, resulting in a landslide that threatened to destroy a community below. Although Skyline lost the bond money, the state paid $400,000 to control the slide. Clearly, unscrupulous mining companies must be forced into responsibility by adequate bond requirements.

One topic of current debate is mountaintop removal, an increasingly popular alternative to strip mining in Appalachia. This method of mining removes the top of a mountain and deposits the overburden into the surrounding valleys and hollows, resulting in a large plateau called a hollow fill. Mining companies find that the plateaus produced by this method are more valuable for farmland and other developments than the original mountain terrain. Critics object to more than the aesthetic presumption of turning a mountain range into flat-topped mesas. Engineers worry that hollow fills may not be stable enough to withstand the heavy rains common to Appalachia; if a hollow fill began to erode, it would result in a landslide of monumental proportions.

Fortunately, environmental groups are trying to push the government to discover and punish violators of SMCRA as well as to raise current standards for environmentally sound strip mining. However, concerned citizens need to be on the alert for potential changes in strip mining regulations and enforcements that could damage the environment. For instance, in 1989 President Reagan proposed opening Appalachian national parks like the Smokey Mountains to strip mining. Only continual pressure on our federal and state representatives will reduce the environmental threat of strip mining to an acceptable level.

SUGGESTED READING

Annual Outlook for U.S. Coal 1988. Energy Information Administration.

Branson, Branley Allan. "Is There Life After Strip Mining?" *Natural History*. August 1986. 31-36.

Caudill, Harry M. *My Land is Dying*. New York: E.P. Dutton, 1971.

———. *Night Comes to the Cumberlands: A Biography of a Depressed Area*. Boston: Little, Brown/Atlantic Monthly Press, 1962.

Conrad, Jim. "From Forests to Forage." *American Forests*. December 1983. 38+.

Doyle, William S. *Strip Mining of Coal: Environmental Solutions*. Park Ridge, NJ: Noyes Data, 1976.

Ergood, Bruce, and Bruce E. Kuhre, eds. *Appalachia: Social Context Past and Present, Third Edition*. Dubuque, IA: Kendall/Hunt, 1991.

Giardina, Denise. "A Coal Field Victory" *The Progressive*. March 1989. 14-15.

Gillis, Anna Maria. "Bringing Back the Land" *Bioscience*. February 1991. 68-71.

Hall, Bob, and Mary Lee Kerr. *1991-1992 Green Index: A State-by-State Guide to the Nation's Environmental Health*. Washington, DC: Island Press, 1991.

Harvey, Curtis E. *Coal in Appalachia: An Economic Analysis*. Lexington, KY: University Press of Kentucky, 1986.

Key, Marcus M., Lorin E. Kerr, and Merle Bundy. *Pulmonary Reactions to Coal Dust: A Review of U.S. Experience*. New York: Academic Press, 1971.

Landy, Marc Karnis. *The Politics of Environmental Reform: Controlling Kentucky Strip Mining*. Washington, DC: Resources for the Future, 1976.

Levy, Builder. *Images of Appalachian Coal Fields*. Philadelphia: Temple University Press, 1989.

Marx, Wesley. "Can Strip Mining Clean Up Its Act?" *Reader's Digest*. March 1987. 121-125.

Mitchell, John G. "The Mountains, the Miners, and Mister Caudill." *Audubon*. November 1988. 80-102.

Morgan, Mark L. *The Enforcement of Strip Mining Laws in Three Appalachian States: Kentucky, West Virginia, and Pennsylvania*. Washington, DC: Center for Science in the Public Interest, 1975.

Munn, Robert F. *Strip Mining: An Annotated Bibliography*. Morgantown: WVU Library, 1973.

Randall, Alan, et al. "Reclaiming Coal Surface Mines in Central Appalachia: A Case Study of the Benefits and Costs." *Land Economics*. November 1978. 472-489.

Rowe, James E., ed. *Coal Surface Mining: Impacts of Reclamation*. Boulder, CO: Westview Press, 1979.

Shapiro, Henry. *Appalachia on Our Minds*. Chapel Hill: University of North Carolina Press, 1978.

Stacks, John F. *Stripping*. San Francisco: Sierra Club, 1972.

Tompkins, Dorothy. *Strip Mining for Coal*. Berkeley CA, : Institute of Governmental Studies, University of California, 1973.

United States Congress. House Committee on Government Operations. Environment, Energy, and Natural Resources Subcommittee. *Strip Mining and Flooding in Appalachia*: Hearings, 95th Congress, 1st Session, July 26, 1977. Washington, DC: U.S. Government Printing Office, 1977.

U.S. General Accounting Office. *Surface Mining: Difficulties in Reclaiming Mined Lands in Pennsylvania and West Virginia*. Washington, DC: General Accounting Office, 1986.

Vietor, Richard H.K. *Environmental Politics and the Coal Coalition*. College Station, TX: Texas A & M University Press, 1980.

Weis, Ina J. *Strip Mining, Environmental Aspects: A Bibliography*. Monticello, IL: Vance Bibliographies, 1982.

CHAPTER *10*

ALTERNATIVE AND RENEWABLE ENERGY RESOURCES: COLUMBUS'S TRASH-BURNING POWER PLANT

Although our society is becoming increasingly aware of the costs and dangers of our reliance on fossil fuels, many of us feel uncertain about what alternatives we really have. One very effective way we can reduce consumption of fossil fuels is to increase the efficiency with which we use energy by conserving, recycling, and adopting new, energy-saving technologies. Individuals and industry alike can significantly conserve energy simply by reducing waste; for example, the amount of heat that leaks out of American windows and doors each year is equivalent to the amount of energy that flows through the Alaskan pipeline annually. In addition, recycling a beverage can or a daily edition of the *New York Times* or *Washington Post* saves the energy equivalent of the gasoline it would take to fill half that beverage can. New compact fluorescent light bulbs can achieve astounding rates of efficiency when combined with lighting systems capable of dimming according to the available daylight or of automatically turning off and on depending on the presence of someone in the room. Nationwide use of these superefficient lighting systems could save about $30 billion worth of utility fuel and lamp replacement each year and cut the nation's total electricity use by 20-25 percent.

Although using energy more efficiently can greatly decrease our reliance on fossil fuels, it cannot eliminate our need for energy altogether. In fact, the expected growth of our population means that we may need more energy than ever in the future even if we successfully implement efficiency measures. Therefore, we must explore other sources of energy. Many environmentalists believe we should devote more efforts to exploring the potential of renewable energy sources such as solar, wind, tidal, hydroelectric, geothermal (heat generated by natural processes occurring beneath the earth's surface), biomass (products derived directly or indirectly from plant photosynthesis such as wood and animal wastes), and trash conversion.

Alternative energy sources are not new to humankind. For centuries, people have used solar radiation from the sun and heat from the earth and have harnessed the power of winds, water, and ocean tides. The ancient Persians used water wheels for turning and grinding; windmills have been a regular power source in Holland and were also used by farmers and ranchers in the United States to operate water pumps; and hot water from thermal springs heated pools used by Romans when they ruled England.

In a modern society, however, there are limitations associated with renewable resources. Most are site specific; that is, they cannot be transported from one area or nation to another, as oil and coal can. Moreover, solar, wind, hydroelectric, and geothermal power are constantly changing or variable; if we are to take advantage of these resources, we must be able to adapt or compensate for their variability.

But alternative fuels also offer many advantages. They can be used to supplement, and eventually replace, fuels that are dwindling or whose cost is or will be prohibitive in the future. According to the Solar Energy Research Institute, renewable sources can meet as much as one-third of the U.S. energy demands by the year 2000. Using alternative sources will broaden our fuel base, allowing us to avoid dependence on any one fuel or on foreign suppliers. And although they are not perfect fuels, alternatives are generally less polluting than fossil fuels.

The idea of incinerating refuse and using the resultant heat to produce electricity is one that gained popularity during the oil crisis of the 1970s. However, with the growing concern over shrinking landfill space for garbage, these incinerators are attracting renewed attention for their ability to reduce the mass of refuse we must dispose of. The trash-burning power plant in Columbus, Ohio, was originally proposed in 1973 in response to the city's growing amount of refuse, and its history reveals many of the benefits, problems, and concerns surrounding such facilities. In addition, the plant does produce electricity and thus raises some of the issues related to the development and use of alternative energy sources.

In developing and using the plant, Columbus has faced a host of problems ranging from potentially toxic waste to economic inefficiency. Following major renovation and technological developments, however, the plant is becoming more and more successful. Although many environmentalists are still concerned about air pollution and toxic waste from refuse burning, most agree that, if combined with extensive recycling, these plants could become environmentally sound ways to reclaim some of the energy invested in what we normally throw away as garbage.

PHYSICAL BOUNDARIES: THE NATIONAL GARBAGE CRISIS AND THE INCINERATION FACILITY IN COLUMBUS, OHIO

As of 1989, over 100 cities in the United States were operating plants that incinerated refuse, and plans existed that would double that number by 1992. Most of these plants, like the facility at Columbus, Ohio, use the resulting energy to produce electricity, but the driving force behind their construction has been what some would call the national garbage crisis. The amount of garbage thrown away in the United States reached 227.1 million tons in 1989. Although the amount of refuse produced per person has remained fairly constant over the years, a growing population has caused the total mass of garbage to rise over time. The Environmental Protection Agency expects this trend to continue and predicts that waste generation in the United States will increase by roughly 20 percent in the next decade.

At the same time that the amount of trash we produce has grown, the wisdom of our traditional method of garbage disposal has come into question. As of 1990, approximately 73 percent of municipal solid wastes was disposed of in landfills; only 13 percent was recycled and 14 percent incinerated. However, landfills all over the country have been closing at increasing rates, either because they are full or because they have been deemed unsafe (toxic substances in landfills can easily wash into the surrounding soil and water supplies). Almost 3000 landfills closed in the second half of the 1980s alone. In addition, the growing awareness of the potential danger that landfills pose both to human health and to the environment is making it more and more difficult to open new landfills. As a result, communities across the country are faced with the need to drastically reduce the amount of garbage they must bury in landfills in a very short time.

Incineration facilities like the one in Columbus, Ohio, are one possibility many communities are examining. Located just south of downtown Columbus in Franklin County, Ohio, the plant occupies a 52-acre site. This site contains the main facility where refuse and coal are burned, a refuse shredder station, a warehouse, water and waste treatment facilities, and a circulating water pump house. The plant also leases a 180 acre lake adjacent to its property; water from the lake is used to collect ash and to dampen coal to reduce dust.

When burned in the main facility, the refuse and coal heat water in boilers above them. Steam produced by the boiling water then powers turbines, which produce electricity. Designers originally predicted that the plant would be capable of burning an average of 2000 tons of refuse per day and generating 90 megawatts of electricity. As of 1989, the plant served approximately 10,000 private and commercial customers in the greater metropolitan area as well as providing street and expressway lighting for areas surrounding the city. In 1991, the plant burned an average of 1750 tons per day and produced about 40 megawatts of electricity per day.

BIOLOGICAL BOUNDARIES: THE POTENTIAL HAZARDS OF INCINERATION

Potentially, incineration plants such as the Columbus Solid Waste Reduction Facility could adversely affect the environment and human health in several ways. The ash that is left after the refuse is burned is usually sent to a landfill. However, if it contains toxic substances it should be sent to a special landfill for toxic waste that prevents anything dangerous from leaking into the environment. If contaminated ash goes to a normal landfill, the toxic substances could enter the groundwater and spread throughout the area. The two toxic substances most commonly detected in incinerator ash are lead and cadmium. Cadmium is a potential carcinogen and lead can harm the central nervous system, blood-forming tissues, the liver, the kidney, and the gastrointestinal system.

Incineration facilities must also remove toxic waste from the water used to wash ash out of the boiler. The Columbus plant includes a water treatment facility to clean its water; the resulting waste goes to its waste treatment facility. If the plant failed to adequately process the water, these pollutants could enter the groundwater or local streams.

Trash-burning plants can also pose the problem of toxic air-born emissions, usually because of metal and chlorine concentrations in the refuse and the sulfur content of the coal. Refuse incineration typically produces particulates (solid matter to which other pollutants adhere), dioxin, furan, and acid-forming gases such as sulfur dioxide, carbon monoxide, and nitrogen oxide. Dioxin and furan are organic compounds suspected of causing birth defects and cancer, and the acid-forming gases can lead to the production of acid rain. Acid rain has been linked to the deterioration of forests, lakes and air quality, especially in the northeast area of the United States, where wind patterns regularly bring in acid-forming gases produced in the midwest.

SOCIAL BOUNDARIES: THE CULTURAL FORCES BEHIND INCINERATION

The greater Columbus area occupies about 187 square miles, with a population of approximately 1,500,000 people. One of the fastest growing cities in the United States, Columbus has experienced many "growing pains," including the dilemma of what to do with a seemingly endless supply of refuse. Garbage production in Franklin county increased by almost 400 percent in the thirty years between 1953 and 1983. In the early 1970s, city officials realized that central Ohio landfills would not hold all of the refuse that Columbus and its outlying areas would produce in the next decade. Studies that revealed the environmental drawbacks of landfills simply added to their concern, and the trash-burning power plant was conceived as a solution to this dilemma.

Although waste reduction was the prime motivation, another social factor behind Columbus's decision to build a refuse derived fuel (RDF) plant was the rising cost and dwindling reserves of nonrenewable energy sources. As oil reserves dwindle and society relies more and more heavily on electricity—which is usually produced by burning coal—communities like Columbus have come to realize that they need to develop alternate energy sources like geothermal power and burning wood and refuse. The only way that communities can ensure their continued growth and success is by long-range planning for energy that does not depend on nonrenewable resources.

LOOKING BACK: THE HISTORY OF INCINERATION
AND THE COLUMBUS TRASH-BURNING PLANT

Although many people today see refuse-derived fuel plants as important sources of alternative energy, people first thought of burning refuse simply as a way to reduce the amount of trash they had to dispose of. The first known operation that burned refuse and then used the resulting heat to produce electricity was built in Oldham, England, in 1896. Although this "plant" was small and crude, it was the forerunner of contemporary facilities.

Most of our current refuse-burning facilities can be divided into two types. Approximately two-thirds of the facilities in the United States practice mass burning, the incineration of mass quantities of unsorted trash at very high temperatures. Mass burn facilities often simply reduce the volume and weight of the trash to be landfilled; those that also generate steam and/or electricity do so as a second priority.

Plants that incinerate refuse-derived fuel (RDF), like the Columbus plant, more effectively combine waste disposal with recycling and energy production. RDF operations allow the removal of recyclable and nonburnable items before incineration. Most RDF plants burn coal as well as refuse to ensure the steady production of electricity and to prevent corrosion in the pipes. Coal combustion produces sulfur oxide, which neutralizes the corrosive effect of chlorine gas, a potential byproduct of burning refuse. The ratio of refuse to coal differs from plant to plant.

The 1960s and early 1970s witnessed the construction of RDF plants in many countries, including the United States, Austria, Switzerland, France, the Netherlands, and West Germany. The city of Columbus saw the relative success of most of these plants, and in 1973 proposed an RDF facility as a solution to the problem of solid waste. The ongoing energy crisis made the idea seem even more attractive. Although Franklin County voters voted down a bond issue to fund the building of the plant in 1976, the idea was not forgotten. The Columbus Council of South Side Organizations supported the issue and, following an intense public education program, it was placed on the ballot once again in 1977, when it passed by a margin of over 60 percent. Construction began in July 1979.

Designing an RDF Plant for Columbus

In designing the plant, engineers had to take into consideration the environmental, health and safety hazards of burning refuse as well as the goal of efficient operation. The plant included storage facilities for up to 140 tons of refuse, and all refuse had to pass through a shredder before being incinerated. Because volatile gases from the refuse can build up in storage areas and shredders, explosions were one hazard designers had to guard against. They decided to surround the shredders at the Columbus plant with heavy concrete walls designed to withstand explosions and to build the shredders without roofs so that pressure from volatile gases could not accumulate as easily. Another potential hazard for plant workers was the airborne pollutants contained in coal dust. To minimize workers' exposure to this dust, the plant was designed so that water could be sprayed on the coal to control dust.

Refuse-derived fuel plants burn fuel at much lower heats than do mass burn facilities. Consequently, they cannot burn items containing metal and steel, which account for approximately 3 percent by weight of incoming municipal solid waste. In the Columbus plant, designers included a magnetic separator to remove these objects after they pass through the shredder. Metal separation is supervised by a private company, which recycles the metallic objects recovered.

The Columbus plant uses an electrostatic precipitator to remove acid-forming gases from the plant's emissions. The use of low-sulfur coal (less than 0.9 percent sulfur) also reduces the amount of sulfur and acid-forming gases produced. Finally, each of the plant's three emission stacks contains an Environmental Protection Agency monitoring device to check emissions for unsafe levels of pollutants.

Making the Plant Work

Although the plant was ready for use in 1983, the city went on to face a number of problems relating to its efficiency and safety. The projected operating efficiency of the plant was drastically reduced by problems with the plant's fuel-feeding, fuel-processing, and ash-handling systems. Because of faulty design, the ash handling system often broke down, allowing as much as a foot of waterlogged ash to pile up on the plant's 1 acre floor. Workers spent up to 75 percent of their time cleaning up and repairing the ash system. The shredders failed to remove such large items as stop signs, tires, and car hoods, which jammed the system and prevented efficient burning. Log jams of refuse also clogged up the system, and the build-up of clinker, the noncombustible material that remains after fuel is burnt, blocked air flow into the cooling system and caused the boilers to overheat. Because of all these problems, in 1984, the plant processed an average of only 772 tons of refuse per day, less than half of the 2000 tons per day originally predicted.

The future of the "cash-burning plant," as it came to be called, became a local controversy. Some people wanted the city to cut its losses by closing the plant. However, in 1984, newly elected mayor Dana Rinehart commissioned experts from The Ohio State University and Battelle Research Institute to study ways the plant might be modified to run efficiently. These experts, along with a task force of city officials, came up with a comprehensive plan to rehabilitate the plant known as "The Big Fix."

Costing taxpayers around $12 million, "The Big Fix" focused on repairing all the major problems contributing to the plant's inefficiency. The Big Fix team and plant personnel redesigned the entire ash handling system and installed stronger magnetic separators to remove all iron- and steel-containing material. They also improved the removal of large refuse items for re-shredding. The fuel feeding system was refitted with vibrating conveyor belts, which feed fuel in a smooth, even stream and eliminate jams. The team also decided the plant should burn coal and refuse separately to avoid the build-up of clinker.

The plant experienced problems with environmental safety as well. During the fall of 1985, a Franklin county landfill refused to accept ash from the plant because it contained unsafe amounts of lead and cadmium. While Battelle and two independent laboratories performed tests on ash samples, the plant was forced to burn only coal. The tests confirmed unsafe levels of lead, although cadmium levels were safe. The plant attempted to reduce the amount of lead in its ash by refusing to accept commercial garbage and asking residential customers not to throw away sources of lead, such as used motor oil and flashlight and car batteries. Currently, the plant mixes its ash with a compound called Alka-Ment. This compound prevents organic acids in landfills from dissolving lead present in ash, thus lessening the possibility that lead could leach out of the landfill.

Ongoing Efforts to Meet the Community's Needs

The city's efforts to improve the plant's efficiency were largely successful. As of 1987, the plant was handling an average of 1600 tons per day, although it burnt anywhere from 2200 to 2400 tons per day on many days. By 1988, it was burning approximately 70 percent of the trash produced in Franklin County. The technological improvements of the Big Fix also improved the plant's economic efficiency. As it burns more refuse, it produces more electricity; electricity generation at the plant rose 5 percent between 1986 and 1987. In that same year, the plant increased the percentage of refuse that it burned, using 52 percent less coal and saving an average of $6000 a day.

In 1989, however, it became apparent that the push for the plant to produce more electricity was having unwelcome side effects: important inspections and repairs had been delayed so as not to interfere with operations, and the practice of mixing coal with the trash to provide more steam was causing boilers to deteriorate more rapidly than expected. Problems due to lack of maintenance caused the average amount of refuse burned per day to drop from 1759 in 1988 to 1600 in 1989.

At the same time, county residents became more aware than ever of the plant's importance. The county landfill was predicted to have space for the residents' refuse for the next 10 years, but only if the plant continued to function; without the plant, landfill space was expected to last for only three more years. Because pressure to produce electricity seemed to be interfering with the plant's ability to incinerate refuse, in 1989 the city officially changed the plant's primary mission to that of trash disposal. Accordingly, in 1990 the plant's name was changed from the Refuse and Coal Fired Municipal Electric Plant to the Solid Waste Reduction Facility. In addition, Mayor Rinehart proposed a $94 million plan to repair, monitor, and expand the plant that would enable it keep up with growth in the county until the year 2010.

LOOKING AHEAD: THE FUTURE OF INCINERATION AND THE COLUMBUS TRASH-BURNING PLANT

Although great strides have been made to improve the efficiency and safety of the Columbus plant, it must go through further changes before it becomes completely economically self-sufficient and environmentally sound. A new Ohio law requires Franklin County to reduce the amount of trash it landfills by 25 percent by 1999; thus, solving the plant's remaining problems has become more essential than ever. Many of the issues facing the plant in the future represent those facing similar waste to energy plants across the country.

Improving Efficiency

One of the most pressing problems facing the plant today is its lack of economic viability. Currently, the plant requires approximately $20 million per year in subsidies from the city of Columbus. Unfortunately, the plant cannot generate more income by raising its tipping fees— the fees it charges private garbage haulers to "tip" or dispose of their loads— because then its fees would no longer be competitive with those charged by the Franklin County Landfill. The landfill cannot raise its fees (which would allow the plant to raise fees as well) because its expenses do not warrant it.

Some claim that transferring the plant to private ownership might help keep losses down. They cite an RDF plant in Akron, Ohio, which cut losses from $2.5 million per year to less than $500,000 per year under private management. However, such a transfer seems unlikely in the case of the Columbus plant. Instead, the city is considering transferring responsibility for the plant to the Franklin County Solid Waste Management Authority (SWMA). SWMA currently operates the county landfill, and if it operated the plant as well, it could legitimately raise tipping fees at both sites.

However, negotiations to transfer the plant to SWMA management were still incomplete as of 1992; both the city and SWMA were concerned about the money the city had invested in the plant and the subsidies that may still be required in the future.

Another step that would improve the plant's operating efficiency would be to institute front end separation, a process that removes nonferrous metals such as aluminum from trash before it is incinerated. Currently, such metals simply melt during the incineration process and their presence requites increased levels of maintenance. In addition, front end separation could help eliminate flammable and combustible materials from the waste stream. Currently, the trash is presorted by hand for such materials, but explosions at shredding plants due to flammables and combustibles are a recurring problem. For example, three out of the plant's five shredding station were inoperable during May 1990 because of explosions.

RDF plants almost always face a certain lack of efficiency in energy production. Because the amount, type, and average size of the refuse they burn varies from day to day, they have difficulty maintaining a steady fire or a steady output of electricity. If the plant cannot produce electricity at a steady rate, it cannot compete with coal-burning plants that produce consistent amounts of electricity and therefore can charge lower rates. If the plant cannot compete in selling its electricity, it must be heavily subsidized. In that case, the taxpayers may decide that the environmental benefits of recycling and burning refuse do not make up for the cost of the plant.

In an attempt to combat this problem, in 1990 the Columbus plant began a pilot project to burn a mixture of natural gas and methane (obtained from old landfills) with its refuse. Plant officials believe natural gas will regulate flame and temperatures more effectively than coal; both refuse and coal take a long time to heat up and to cool down, but natural gas heat can be adjusted almost instantaneously. They hope this project will allow them to compete more effectively with coal burning plants and eventually make their plant a profitable enterprise.

Despite the overwhelming amount of refuse that prompted Columbus to build the plant, an occasional temporary *lack* of garbage can also cause the plant to operate less efficiently. For instance, in January 1985, Columbus had to suspend trash collection because of sub-zero temperatures, and thus the plant had to rely totally on coal for fuel. Although they managed to maintain customer service, it was much more expensive to provide the normal amount of electricity. Although such incidents cannot be prevented, an increasing supply of refuse should help offset any temporary holdups in the fuel coming in to the plant.

Protecting the Environment

Although the Columbus plant solved its toxic ash problem by adjusting the amount of lead in incoming refuse and using Alka-Ment, many environmentalists worry that RDF plants threaten the environment because of unenforced regulations and unregulated emissions.

Environmental Protection Agency tests on ash from RDF plants all over the country show that fly ash, the fine particulate matter trapped in air-control devices, almost always shows unacceptable levels of toxic metals like lead and cadmium. However, bottom ash, which is washed from underneath the boilers, is unacceptable only 10-30 percent of the time. Although plants make ash acceptable for disposal by mixing fly and bottom ash, the composition of the refuse burned varies from day to day. To determine if ash is safe for regular landfills, where toxics could leach into groundwater, ash should be tested every day. However, many plants don't want the extra expense and effort of daily testing, and they also want to avoid using toxic waste landfills, which can cost up to 15 times as much as regular landfills. Because the Environmental Protection Agency does not insist upon daily testing, some experts fear undetected toxic waste from ash is contaminating our environment.

Environmentalists also point out that, as of 1987, only lead, mercury, and beryllium out of the 27 toxic metals that refuse-burning might produce were controlled by government regulations. Dioxin and furan, although suspected carcinogens, were not regulated by the federal government at all. Although the EPA reports the risk of cancer from incinerator emissions to be small, critics claim the scope of their studies and the quality of their testing is insufficient. Overall, such critics claim that regulations affecting emissions need to be much more thorough and more strictly enforced if the ever growing number of trash-burning plants in the country are not to seriously damage air quality.

Although acid-forming emissions from municipal waste combustion facilities will be affected by the Clean Air Act reauthorized in 1991, standards have not yet been established for such facilities. Experts predict, though, that plants like the one in Columbus may be required to install scrubbers or other costly pollution control devices. Such measures could cost the Columbus plant up to $32 million by 1996.

Environmentalists are also concerned about the relationship between RDF plants and recycling. Although RDF plants can remove many recyclable objects before incineration, not all plants recycle as much as they can. By burning recyclables, they may undercut other recycling efforts in their areas even though studies have shown that recycling is often more economical than burning or landfilling. For example, burning 1 pound of paper produces 500 BTUs of steam, whereas recycling that same pound would conserve 2000 BTUs in energy required to produce paper from virgin pulp. In addition, recyclable items like glass and paper add to the mass of ash that must be sent to a landfill after incineration. Finally, these items can also increase the plant's level of toxic emissions. By instituting front-end separation, the Columbus plant would allow easily recycled materials such as glass and aluminum to be removed before incineration, thus diminishing its negative impact on recycling efforts.

Some studies predict that, by 2001, our nation will process half of its trash in some way, including RDF plants, rather than sending it to landfills. Given this trend, RDF plants could play an important role in solving world problems with refuse disposal and energy needs; with improved technology, plants like the Columbus Solid Waste Reduction Facility have the potential to be environmentally sound as well as economically efficient. But to realize the full potential of RDF plants, we must make sure that potentially toxic ash and airborne emissions are strictly monitored and controlled. RDF plants must also strive to recycle as much as possible in order to provide the maximum benefit to the environment and to the societies that depend on it.

SUGGESTED READING

Boegly, William, Jr. *Solid Waste Incineration—Incineration with Heat Recovery.* Springfield, VA: Oak Ridge National Laboratory. 1978.

Cook, Herb, Jr. "Trash: A Critical Question for Our Times." *Columbus Monthly.* March 1991. 48-54.

Flavin, Christopher, and Rick Piltz. *Sustainable Energy.* Washington, DC: Renew America. 1989.

Flavin, Christopher, and Nicholas Lenssen. "Designing a Sustainable Energy System." *State of the World 1991.* New York: W.W. Norton and Company, 1991. 21-38.

Kirschner, Dan. *To Burn or Not To Burn: The Economic Advantages of Recycling over Garbage Incineration for New York City.* New York, NY: Environmental Defense Fund, 1985.

Lovins, Amory. *Soft Energy Paths: Toward a Durable Peace.* Cambridge, MA: Ballinger Publishing Company, 1977.

Mann, Carolyn. "Garbage In, Garbage Out." *Sierra.* September/October 1987. 21-32.

National Center for Resource Recovery. *Incineration: A State of the Art Study.* Lexington, MA: Lexington Books, 1974.

Neal, A.W. *Refuse Recycling and Incineration*. Stonehouse: Technicopy Limited, 1979.

Newsday, Inc. *Rush to Burn: Solving America's Garbage Crisis?* Washington, DC: Island Press, 1989.

Rochford, R.S., and S.J. Witkowski. *Considerations in the Design of a Shredded Municipal Refuse Burning and Heat Recovery System*. New York, 1978.

Rubel, Fred N. *Incineration of Solid Wastes*. Park Ridge, NJ: Noyes Data Corporation, 1974.

Todd, Gordon. "City RDF Plant Stops Burning Cash." *Waste Age*. November 1987. 50-58.

Vaughan, D.A., H.H. Krause, and W.K. Boyd. *Handling and Co-Firing of Shredded Municipal Refuse and Coal in Shredder-Stoker Boiler*. New York, 1978.

Weinstein, Norman J., and Richard F. Toro. *Thermal Processing of Municipal Solid Waste for Resource and Energy Recovery*. Ann Arbor, MI: Ann Arbor Science Publishers, 1976.

Young, John E. "Reducing Waste, Saving Materials." *State of the World 1991*. New York: W.W. Norton and Company, 1991. 39-55.

CHAPTER *11*

AIR RESOURCES: ACID PRECIPITATION IN THE ADIRONDACK MOUNTAINS

Of all the elements that are essential for life, air is truly the most global, disseminating around the world more easily even than water. Unfortunately, this quality also makes cleaning up and controlling air pollution an especially difficult problem. The various pollutants entering our atmosphere—many of them produced by the burning of fossil fuels such as oil and coal— can rarely be traced to a single source, although some countries, cities, and even specific factories or utilities can be identified as particularly serious sources of pollution.

Because of their global nature, it is not surprising that problems stemming from the pollution of air resources, such as global warming (commonly known as the greenhouse effect) and the thinning of the ozone layer, are some of the most pressing and well-known environmental issues today. Just as infamous, acid precipitation, commonly called acid rain, also results from air pollution and provides an especially good example of the way one region's pollution can become another region's problem. In this case study, we will look closely at the Adirondack Mountains, one of the first regions in the United States to suffer significant and widespread damage from acid precipitation (which includes acidic snow, mist, and fog as well as rain).

Once renowned for their unspoiled wilderness areas and hundreds of lakes teeming with fish, the Adirondacks have achieved notoriety of a more dubious nature. Many of the most beautiful lakes have gained their clear, sparkling waters at the cost of their living animal and plant matter, just one result of continued assault by acid precipitation.

PHYSICAL BOUNDARIES: HOW ACID PRECIPITATION OCCURS AND WHY THE ADIRONDACKS ARE VULNERABLE

In order to evaluate the threat that acid precipitation poses to the Adirondack region, it is important to understand both the mechanism by which acid precipitation is formed and the factors that make an area like the Adirondacks especially vulnerable to damage from that precipitation.

The Formation of Acid Precipitation

Although normal precipitation is slightly acidic, it is not harmful to plant or animal life. In contrast, the increased acidity of acid precipitation can pose a threat to living organisms and ecosystems as well as to buildings, monuments, and other material objects. To determine the acidity of rain or snow, we use a pH scale that measures acidity or alkalinity along a continuum from 1 (a strong acid like battery acid) to 14 (a strong alkali or base like lye). Because the pH scale is logarithmic, each change in whole numbers represents a tenfold increase or decrease in acidity or alkalinity. For example, a pH of 4 contains ten times as much acid as a pH of 5, and a hundred times more acid than a pH of 6. Normal precipitation might have a pH as low as 5.6 to 5.9. However, a pH reading of 5.5 or lower may indicate acidic precipitation and is thus a cause for concern.

Acid precipitation is created when certain chemical compounds react with the water vapor naturally present in the air. The two major compounds involved in this reaction, also known as precursors of acid precipitation, are sulfur dioxide (SO_2) and nitrogen oxides (NO_x). Although the amount of SO_2 and NO_x from natural sources exceeds that produced by human activities, or cultural sources, naturally produced SO_2 and NO_x can usually be removed from the atmosphere by natural processes. However, the additional amount produced by cultural sources can exceed nature's recycling capacity, thus remaining in the atmosphere and eventually forming acid precipitation.

The major cultural sources of SO_2 are electric generating utilities, manufacturing plants powered by fossil fuels (especially coal and petroleum), smelters that process sulfur ores, and the production of sulfuric acid for industrial purposes. In the northeastern United States, between 75 and 80 percent of acid precipitation can be traced to sulfur oxides. In the west, however, nitrogen oxides are the chief culprits in the formation of acid precipitation. The largest single cultural source of nitrogen oxides is internal combustion engines that power automobiles, buses, and other vehicles. Stationary sources like electric power plants, manufacturing plants that produce high grade metals, home heating units, and incinerators account for about half the total load of culturally produced NO_x in the atmosphere.

Once emitted into the atmosphere, sulfur dioxide transforms into sulfate particles and nitrogen oxides into nitrate particles. Precipitation, or wet deposition, of acids occurs when the sulfate or nitrate particles react with water vapor to form sulfuric acid and nitric acids, which then fall to the earth in rain, snow, mist, or fog. Sometimes the sulfates or nitrates fall to the earth as dry deposition, then combine with water in a stream, lake, or pond to form acids.

Why the Adirondacks are Sensitive to Acid Precipitation

The Adirondack Mountains stretch from the Mohawk River Valley in the south to the St. Lawrence River in the north, and extend west from Lake Champlain to the Black River. Although located near the heavily populated eastern seaboard, the Adirondacks encompass twelve wilderness areas so remote that they can be reached only by canoe or by foot and a northwest stretch of 50,000 acres that has never been touched by loggers. So much of the Adirondacks has remained wild because six million acres of the mountains are in the Adirondack State Park. The largest park in the lower 48 states, it contains more land than the state of Massachusetts. Unfortunately, this vast natural preserve is especially vulnerable to the effects of acid precipitation because of its location, altitude, and geological formation.

The northeastern region of the United States is subject to an unusually large amount of acid-forming emissions; about 90 percent of the sulfur, nitrogen, and other acid precursors found in the northeast United States result from human activities. Approximately 75 percent of the acid deposition in the Adirondacks originates in the midwest, where electric power plants often burn high-sulfur coal. Prevailing air currents transport SO_2 and NO_x from these plants hundreds of miles eastward. The long journey allows ample time for the SO_2 and NO_x to form sulfuric and nitric acids. After three days aloft, as much as 50 percent of available SO_2 converts to acid precipitation.

When air masses reach the Adirondack Mountains, the first high-altitude area in the eastward path, they tend to ascend and cool, often resulting in rain or snow. Thus, air currents regularly drop the bulk of their acid load on the Adirondacks as they move to the east. Because the higher peaks receive more acid deposition, lakes and vegetation in the more remote, high-altitude regions of the Adirondacks were the first to show adverse effects.

The Adirondacks are also sensitive to acid deposition because of their geological structure. As an extension of the Canadian Shield, a large sheet of Pre-Cambrian granite, they lack the thick soil cover that could neutralize at least some of the excessive acids from the precipitation. Consequently, the acids can more easily reach vegetation and lakes.

BIOLOGICAL BOUNDARIES: HOW ACID PRECIPITATION AFFECTS LIVING ORGANISMS AND SYSTEMS

As a state park, the Adirondacks have provided a safe, natural haven for a wide variety of plants and animals. The forests contain large numbers of northern hardwoods, hemlock, and white pine. Conifers, especially spruce and balsam, flourish at higher elevations. The 2200 ponds and lakes are home to a variety of sporting fish, including trout and salmon, as well as other aquatic life. Deer, black bears, small game animals, and coyotes range through the forests, and a healthy beaver community has re-established itself since it was almost trapped into extinction in the mid-1800s. However, all these populations have become potential victims of acid precipitation.

How Acid Deposition Affects Soil and Vegetation

When acidic rainfall filters through soil, it affects the soil and vegetation as well as the lake it eventually reaches. By bonding with calcium, magnesium, potassium, and other soil nutrients essential to plant growth, acids make these minerals unavailable to vegetation. In the Adirondacks, where soil contains small amounts of these nutrients, this process severely retards the growth of trees and other plants. Acidic precipitation can also dissolve metals from soil and bedrock, thus freeing them and making them available for absorption by plants. Metals such as aluminum and manganese can damage plant roots' ability to absorb nutrients, thus hampering growth, or even killing the roots.

Acid precipitation attacks trees and plants from above as well as from below. SO_2 gases entering trees through stomata on leaves and needles can erode cuticle wax and leave the tree susceptible to insect infestation and loss of moisture. The acid also affects buds and bark and can remove nutrients directly from needles and leaves.

Trees and plants not killed outright may be weakened so much that they are more likely to succumb to natural stresses such as insects, disease, and harsh weather. In the Adirondacks, forests at high altitudes are growing at slower rates and losing leaves and needles. Studies show that red spruce in the Adirondacks are especially at risk, having experienced a substantial decrease in growth rate in the past 25 years. Conifers like the red spruce are the first trees to show damage because they grow at higher altitudes than hardwoods and are thus on the "front line" of attack by acid precipitation. Needles of conifers are also more likely to collect particles and gases from the atmosphere than are the leaves of deciduous trees; thus, they absorb more sulfuric and nitric acid from the acid precipitation they experience.

How Acid Precipitation Affects Water and Aquatic Life

Most lakes receive water not only from direct rainfall, but also from runoff that filters through the watershed, or the soil surrounding the lake. As acidic runoff leaches metals from the soil, it washes them into streams and lakes where they can damage aquatic life. Aluminum is especially deadly for fish because it damages their gills, ruining their ability to absorb oxygen from the water. Other metals, such as mercury, which build up in the systems of fish can be harmful for humans or other animals that eat the fish.

Although runoff picks up toxic metals as it filters through soil, the filtering process can also help to buffer the acids before they reach the lake. Soil rich in minerals like calcium tends to neutralize acid because the minerals bond with the sulfate and nitrate components of the acid. A study published in 1985 by Charles Driscoll of Syracuse University and Robert Newton of Smith College showed that three similar lakes in the Adirondacks which received the same amounts of acid precipitation varied widely in pH. One lake, which retained a neutral pH, was surrounded by soil eight times thicker than the soil around a highly acidified lake. Scientists concluded that the thicker soil gave the acid more opportunity to bond with minerals. However, many plants and animals need these same minerals as nutrients, and if the minerals are being tied up by molecules from acid rain, the plants and animals can suffer. For example, studies of fish in acidified lakes show that a lack of calcium can result in such weak bone structure that their muscles pull their skeletons out of shape.

The acid that remains in groundwater and falls directly into surface water harms aquatic life forms in ways that are still being explored. A study by Winsor Watson and Carl Royce-Malgren of the University of New Hampshire showed that even comparatively low levels of acid (pHs between 5.0 and 5.5) seemed to prevent fish from smelling odors that help them migrate and find mates. Thus, even low levels of acid could virtually end reproduction in entire populations of fish.

If acid precipitation continues to assault a lake, eventually everything in it will be destroyed; almost no forms of aquatic life can live in the resulting high concentrations of acid. Even those that could survive may die from lack of other biota for food. Although estimates vary from study to study, acid precipitation has rendered approximately 200 lakes in the Adirondacks incapable of supporting fish populations. A 1989 study (see Asbury, et al.) compared current acid levels in 274 lakes to measurements taken in a rare study conducted between 1929 and 1934. The results showed that 80 percent of those lakes had acidified in that 50-year period. The ecologists determined acidification by tracing changes in alkalinity, a more accurate method than the pH-based analysis conducted by the National Academy of Sciences (NAS) in 1986; the NAS study had claimed that it was possible no change in acidification had occurred.

How Acid Precipitation Affects the Human Population

The heavy acid precipitation in the Adirondacks also endangers the 120,000 people who live in the area year round as well as the additional 90,000 summer residents. Crops, fish or meat may be contaminated with toxic metals or lacking in nutritional value. High concentrations of SO_2 gas in the air also cause sore throats, coughing, and lung irritation and tissue damage. In addition, the Adirondack region experienced a mysterious outbreak of cases of gastroenteritis in the mid-1980s. Investigators later traced the outbreak to a normally rare form of bacteria. This bacterium was especially resistant to acid, and temporarily flourished when acid precipitation killed off competing life forms.

All water sources in the Adirondacks are in danger of acidification as well, and consuming water with high acid levels is known to increase the chance of cardiovascular disease. An even more immediate problem with acidified water, however, occurs when it runs through lead pipes. As the sulfuric acid corrodes the pipes, the water can become contaminated with lethal concentrations of lead.

SOCIAL BOUNDARIES: THE CULTURAL VALUE OF THE ADIRONDACKS

Because of their proximity to the east coast, the Adirondacks were explored early in the history of the United States. In fact, most of the forest had been burned or cut down at least once by 1885. At this time, people became concerned about the effect of increasing numbers of tourists, settlers, loggers, and other human activity on the beautiful mountain area. The New York State Surveyor Verplanck Golvin had also developed a special interest in the area, which he had explored extensively. His interest and the public concern led to the creation of the Adirondack Forest Preserve in 1885, and the Adirondack Park in 1892. Two years later, a constitutional amendment declared that forest preserve land would be protected as wild forest forever. Today, more than a million people enjoy hiking and camping in the Adirondacks every year.

LOOKING BACK: THE HISTORY OF ACID
PRECIPITATION IN THE ADIRONDACKS

Although few historical records of acid levels in the Adirondacks exist, researchers believe that acid precipitation may have begun to affect the area as early as the beginning of the twentieth century; they attribute the possible increase in acid-forming emissions to the Industrial Revolution of the late 1800s, when industrialization powered by the burning of fossil fuels began to become a way of life in western countries. For instance, Lake Awosting in the Adirondacks had a pH as low as 4.5 in 1930 and was empty of fish life as early as 1915; scientists theorize that the rise in industries like railroads, tanneries, and iron forges in the area during the late 1800s could have been responsible for killing the fish population and acidifying the lake.

On the whole, though, scientists have had to rely on records from the last few decades, since the problem has become obvious. Damage to lakes and trees in the Adirondacks could easily have occurred long before anyone noticed, since the first damage occurs at the highest, most inaccessible altitudes. Acids also tend to kill smaller, weaker fish first, so fishers who only look for the bigger fish would not have noticed any change until the whole population had been affected. However, by the 1970s, people were concerned about obvious changes in lakes and vegetation of the Adirondacks, and scientists began to study why and how these changes were taking place.

The Clean Air Act Controversy

When Congress took a major step to protect air quality in 1963 by passing the Clean Air Act, acid precipitation was a little-known phenomenon and thus the act did little to address acid-causing emissions. However, by the time the Clean Air Act came up for Congressional reauthorization in 1977, acid precipitation had become a focus of controversy. Scientists insisted that acid rain in the Adirondacks and throughout the country would only end when midwestern electric plants curtailed acid-forming emissions. Opponents of emissions controls argued just as vehemently that more study was needed, beginning a debate that would last fourteen years.

In 1984, Representatives Waxman and Sikorski proposed legislation that would require the 50 main plants responsible for acid-forming emissions to install scrubbers, which remove acid-forming compounds before they enter the atmosphere. Although pollution control in the United States has traditionally been based on the philosophy of "the polluter pays," the Waxman-Sikorski bill proposed that the new scrubbers be funded by a tax on fossil fuel energy in all 48 contiguous states. Representatives of the nonpolluting states, especially those from states where utility customers had been footing the bill for "clean" energy for years already, objected. They were opposed by the representatives of the 19 states where those 50 sources are found, especially Pennsylvania, West Virginia, Georgia, Ohio, Indiana, Kentucky, Tennessee, and Missouri, which would be most affected. Elected officials, industry representatives, and the general public in those states maintained that they should not be forced to bear the full costs of cleaning up emissions. They warned that such a move would unfairly target industry in their states and wreak havoc on state economies. Not surprisingly, Congress reached no agreement on the matter.

Other legislation that would have allowed plants to choose their own methods of emissions control also failed. States like Ohio, which produce high-sulfur coal, were afraid that plants would opt for burning low-sulfur coals rather than installing expensive scrubbers, and thus eliminate jobs in their coal industries.

Another way opponents of emissions regulations stalled any decisive action was by insisting on even more research. They claimed that no one knew if expensive measures to control emissions would really reduce acid rain. President Reagan assisted this coalition in 1980 by sponsoring the National Acid Precipitation Act, which called for a ten-year program of research into the causes and consequences of acid precipitation.

In the meantime, further study and attempts at reducing emissions in countries like Great Britain showed a definite correlation between limiting pollution and reducing acid precipitation. Throughout the 1980s, most experts felt that, while more investigation would be helpful in reversing the damage of acid rain, waiting to reduce emissions would only lead to increased and possibly irreversible damage in hard-hit areas like the Adirondacks.

Finally, Congress reauthorized the Clean Air Act in 1990. The act stipulates that SO_2 production must be reduced from today's output of approximately 20 million tons per year to 10 million tons annually by the year 2000. Utility plants affected by this bill must either install scrubbers or switch to low-sulfur coal. In addition, the 14 million tons of NO_x produced annually must be reduced by 2 to 4 million tons per year as soon as 1992.

How Scientists are Attempting to Help

Although the only way to prevent acid rain is to eliminate acid-forming emissions, some scientists are working to make the effects of acid precipitation less harmful to lakes and fish in areas like the Adirondacks. Some researchers are concentrating on the possibility of stocking lakes with fish that are especially good at surviving in acidic waters; geneticists are even trying to breed special strains of acid-resistant fish. However, some environmentalists point out that even if the fish can tolerate acidity, they must also survive high metal concentrations. In addition, they may die from lack of food if the rest of the aquatic population has disappeared. Finally, some believe that successfully finding acid-resistant fish would simply give the "polluter" states and power plants an excuse not to clean up emissions.

A more widely accepted area of experimentation is liming lakes that have high acid levels. Theoretically, the highly alkaline lime is supposed to neutralize the acid. The New York State Department of Environmental Conservation and other private organizations have been conducting studies on liming lakes in the Adirondacks since 1959. They have found that liming usually restores a normal pH, and so can be used in conjunction with restocking programs to restore lakes that have become acidified. Skeptics point out that acid precipitation damages the entire watershed, not just the lake, They emphasize that liming can neither repair watershed damage nor bring back the rest of the biotic life that has been lost. Moreover, liming is expensive. Estimates predict that it would cost 10 to 20 billion dollars to lime several hundred lakes in the Adirondacks over five years; restocking and monitoring efforts would cost even more. Of course, if acid precipitation continues unabated, a limed lake will soon lose its normal pH; some lakes have re-acidified in as few as six months after they were limed.

LOOKING AHEAD: THE FUTURE OF THE ADIRONDACKS AND OTHER AREAS AFFECTED BY ACID PRECIPITATION

The reauthorized Clean Air Act goes a long way toward starting the fight against acid precipitation. In addition to its effect on utilities and other stationary sources of acid precipitation, the act places limits on the amount of NO_x passenger vehicles can produce. However, other measures that could make an important difference remain to be taken; for example, legally raising the minimum number of miles per gallon on new automobiles could drastically cut emissions of NO_x as well as other pollutants such as hydrocarbons. In addition, loopholes in the act actually allow some power plants in the west to increase their emissions of aid precursors and permit utilities presently using "clean coal technologies" to increase emissions of SO2 until 2004, when stricter standards go into effect.

Some scientists have even more serious reservations about the stipulated reductions in emissions. They point out that the National Acid Precipitation Program, which largely supplanted other acid precipitation studies when it was initiated by the federal government in 1980, substantially narrowed the scope and range of acid-induced forest decline studies. Although the amount of forest research improved after 1985, most field studies were too complex to complete before the Clean Air Act was rewritten and passed in 1991. Thus, the standards of the reauthorized act were based largely on the tolerances of lakes and streams rather than those of forests. In addition, the reauthorization eliminated most funding for research just as new data suggesting the important role of nitrogen in forest decline were still being evaluated. According to many scientists, this new data suggest that nitrogen emissions must be reduced much more significantly than stipulated in the act if we are to prevent further damage to forests like those of the Adirondacks.

We must not allow the achievements of the reauthorized Clean Air Act to lull us into a false sense that the problems of acid precipitation are solved. Loopholes must be closed, and new data on acid precipitation's effects on forests should be taken into account. Also, because acid rain is truly a global problem, we must work vigilantly with other countries around the world to help them reduce their production of acid-forming emissions.

SUGGESTED READING

Aldhous, Peter. "Acid Rain: An Economic Case for Environmental Cooperation." *Nature.* May 16, 1991. 175.

Asbury, Clyde E., Frank A. Vertucci, Mark D. Mattson, and Gene E. Likens. "Acidification of Adirondack Lakes" *Environmental Science and Technology.* March 1989. 362-364.

Baker, Lawrence A., Alan T. Herlihy, and Philip K. Kaufmann. "Acid Lakes and Streams in the United States: The Role of Acidic Deposition." *Science.* May 24 1991. 1151-1154.

Barnett, Lincoln Kinnear. *The Ancient Adirondacks.* New York: Time-Life Books, 1974.

Begley, Sharon. "On the Trail of Acid Rain: While Politicians Stall on Solutions, Scientific Sleuths Establish the Link Between Smokestacks and Dead Lakes." *National Wildlife.* February/March 1987. 6-12.

———. "Pollution Knows No Boundaries." *National Wildlife.* February/March 1990. 34-43.

Corcoran, E. "Cleaning Up Coal." *Scientific American.* May 1991. 106-116.

Driscoll, C.T., and G.C. Schefran. "Short-term Changes in the Base Neutralizing Capacity of an Acid Adirondack Lake, New York." *Nature.* July 26 1984. 308-310

Ember,Lois R., David J. Hanson, Janice R. Long, and Pamela S. Zurer. "Congressional Outlook '90." *Chemical and Engineering News.* January 8 1990. 8-13.

Flynn, John. "Forest without Trees: How Congress Was Duped about Acid Rain's Effects." *The Amicus Journal.* Winter 1991. 28-33.

Healy, Bill. *The Adirondacks: A Special World.* Utica, NY: North Country Books, 1986.

Holdren, George R., Thomas M. Brunelle, and Gerald Matisoff. "Timing the Increase in Atmospheric Sulphuric Deposition in the Adirondack Mountains." *Nature*. September 20 1984. 245-248.

Kahaner, Larry. "Something in the Air." *Wilderness*. Winter 1988. 18-27.

Merriam, C. Hart. *The Mammals of the Adirondack Region, Northeastern New York*. New York: Arno Press, 1974.

Raloff, Janet. "Acid Highs and Lows in Adirondack Lakes." *Science News*. March 18 1989. 165.

——. "Acid Rain: Lowdown on Health of Lakes." *Science News*. May 20 1989. 311.

Pearce, Fred. "Whatever Happened to Acid Rain?" *New Scientist*. September 15 1990. 57-60.

"Pouring Forth on Acid Rain." *Environment*. March 1990. 23-24.

Roberts, Leslie. "Acid Rain Progress: Mixed Review." *Science*. April 19 1991. 371.

Schaefer, Paul. *Defending the Wilderness: The Adirondack Writings of Paul Schaefer*. Syracuse, NY: Syracuse University Press, 1989.

Sullivan, T.J., D.F. Charles, and J.P. Smol. "Quantification of Changes in Lakewater Chemistry in Response to Acid Deposition." *Nature*. May 3 1990. 54-58.

Travis, W.B. "Some Truly Hair-Raising Regulations." *Sierra*. January/February 1991. 22.

U.S. Congress Office of Technology Assessment. *Acid Rain and Transported Air Pollutants: Implications for Public Policy*. New York: UNIPUB, 1985.

Van Valkenburgh, Norman J. *The Adirondack Forest Preserve: A Narrative of the Evolution of the Adirondack Forest Preserve of New York State*. Blue Mountain Lake, NY: Adirondack Museum, 1979.

"What You Need to Know about the New Clean Air Act." *EPA Journal*. January/February 1991. 11-60.

Woodin, Sarah, and Ute Skiba. "Liming Fails the Acid Test." *New Scientist*. March 10 1990. 50+.

CHAPTER *12*

WATER RESOURCES: THE WILLAMETTE RIVER BASIN

Although we call our planet "Earth," 71 percent of its surface is covered by oceans and seas. And this figure does not even include all the water held in lakes, ponds, rivers, streams, marshes, bays, and estuaries, as well as the water locked up in polar ice caps and glaciers. Despite the apparent abundance of water, less than 3 percent of the world's water is fresh (the rest is salty). Of that 3 percent, three-quarters is found in polar ice caps and glaciers, and nearly a quarter, known as groundwater, is found underground in water-bearing porous rock and sand, or in gravel formations. Only a small proportion—one-half of 1 percent of all water in the world—is found in lakes, rivers, streams, and the atmosphere.

Unfortunately, human population growth, agricultural practices, and increasing industrialization are placing an ever increasing demand on these water resources even as they pollute more and more of the supply. The Willamette River in Oregon exemplifies many of the problems threatening water resources today. Nearly 70 percent of Oregon's residents live in the basin of the Willamette River, although the basin makes up just 12 percent of the state's area. Early settlers were attracted by the basin's fertile soil, productive forests, and, most importantly, the river itself. Unfortunately, by the early 1900s, overuse and pollution had taken their toll, and many people believed that the Willamette River would never recover.

However, the story of the Willamette also serves as a positive example of the way those who depend on a shared water resource can work together to rejuvenate that body of water and protect it from further harm in the future. As early as the 1920s, citizens of the Willamette Basin made improving the water quality of the river a priority. Because of their efforts, the Willamette Basin is once again a popular and productive region for agriculture, industry, and residential development.

PHYSICAL BOUNDARIES: HOW WEATHER AND SOIL CHARACTERISTICS INFLUENCE THE WILLAMETTE BASIN

The Willamette Basin, roughly 150 miles long and 75 miles wide, is defined by the river and surrounding mountain ranges: the Coast Range on the west, the Cascade Range on the east, and the Calapooya Mountains to the south. The river itself begins in the Calapooyas, at the confluence of the Coast Fork and the Middle Fork. From there it runs 187 miles to empty into the Columbia River near Portland, the northern boundary of the basin. The Willamette is the twelfth largest river in the United States and one of the few major North American rivers to run northward.

When settlers arrived in the 1800s, they naturally flocked to the river because it provided water as well as passage to the Pacific by way of the Columbia River. They were also attracted by the moderate temperatures in the basin, which range from 3° C (38° F) in the winter to 19° C (67° F) in summer. The basin maintains this pleasant range because the Coast Range blocks many of the more violent ocean storms coming from the west. The Cascade Range blocks air of more extreme temperatures from the east.

Although temperature extremes are rare, precipitation extremes are not. Forty-eight percent of the precipitation in the basin falls during the winter months of November, December, and January, whereas only two percent falls in July and August. This variation occurs because of the interaction between ocean air and the land. During the winter, moist warm air comes in from the ocean and rises to cross the Coast Range. As it rises, it cools and forms precipitation, resulting in rain or snow. In the summer, when the temperature of ocean air is cooler than that of land, the air warms as it moves inland; this warming offsets the cooling effect of rising, and the air retains most of its moisture.

The varying rainfall in the basin directly affects the rate at which water flows in the Willamette River. Before the state of Oregon began to regulate flow, the Willamette's rate varied from summer flows as low as 2500 cubic feet per second (cfs) to flood conditions of 500,000 cfs during winter months. While the high-velocity flows of winter and spring made flooding a danger, low flows in the summer impeded agriculture by reducing water available for irrigation.

In addition, low flows cause the river to run more slowly. When river water moves more slowly, sludge and bacteria tend to build up. The bacteria decompose, or break down, organic wastes (small pieces of once living plant or animal matter) found in the sludge in order to obtain energy. Most bacteria are aerobic; in other words, they use oxygen that they find dissolved in the water as they decompose organic matter.

The amount of dissolved oxygen (DO) that bacteria need to decompose the organic matter in water is known as biological oxygen demand (BOD). A body of water with a high BOD will have a low concentration of DO because oxygen is being used up by bacteria to decompose organic matter. Accordingly, a body of water with a high BOD is also high in organic matter. Because lack of DO can cause aquatic organisms such as fish and shellfish to die, oxygen depletion is a major contributor to the degradation, or loss of species diversity, in a body of water.

Thus, the low summer flows of the Willamette acted as a major factor in degrading water quality in the river. Today, controlled releases of water from reservoirs minimize seasonal variation; currently, the yearly average rate of flow at Salem is 23,000 cubic feet per second.

The Willamette River Valley, the 3500 square miles of the basin under 500 feet above sea level, also proved ideal for agriculture. Besides having a ready supply of water, the valley has rich soil, a legacy of the last Ice Age, when a large lake deposited fertile soil in the valley. The floor of the valley is flat and relatively well covered with grasses and trees, and their root systems help prevent soil from eroding away with wind, rains, or melting snow. In the mountains, forest cover also prevents erosion.

BIOLOGICAL BOUNDARIES: USE AND ABUSE OF
NATURAL RESOURCES IN THE WILLAMETTE BASIN

Many of the biological characteristics of the Willamette Basin served as resources for development, agriculture, and industries such as logging and fishing. Ironically, all these activities contributed to the degradation of the Willamette and thus helped to threaten some of the resources that had made their growth possible.

Although forests cover only about 30 percent of the land in the valley (most is used for agriculture or urban development), they cloak nearly three quarters of the total land in the Basin. Conifers dominate; most occur in the mountains and foothills above 1000 feet.

Fish are another important resource of the Willamette Basin; many fish such as salmon and trout migrate from the Pacific Ocean and Columbia River to the Willamette and its tributaries to spawn. The basin is also home to a variety of wildlife, including big-game species like the black bear, Roosevelt elk, and the black-tailed deer. Wintering and migratory birds, such as Canada geese, depend upon the streams, lakes, and reservoirs of the Willamette Basin for resting and feeding grounds.

SOCIAL BOUNDARIES: HUMAN ACTIVITIES AND
THEIR INFLUENCE ON THE WILLAMETTE BASIN

Although explorers and trappers ventured into the Willamette Basin as early as the late 1700s, settlement of the region did not begin in earnest until the 1840s. Before long, the rich natural resources of the area gave rise to a variety of industries. Agriculture quickly became a major source of food and income in the valley. Currently, approximately one third of the basin's 8 million acres are used for crops and grazing. Not surprisingly, a food processing industry also developed in the area, and today there are nearly four hundred of these companies in the basin.

Because of the vast nearby forests, logging also became a major industry in the region. The Willamette and its tributaries were used to transport cut lumber. A ready supply of wood and inexpensive energy generated from the flow of the river gave birth to a thriving pulp and paper industry along the river banks.

Because rivers were such important modes of transportation, several major cities developed along the Willamette as well. Portland, at the junction of the Willamette and Columbia Rivers, is now the largest freshwater port on the Pacific coast and has become a major railroad center. Salem became the state capital of Oregon, while Eugene grew into the dominant center of trade for all of southwest Oregon.

Agriculture, industry, and transportation: each contributed to the economic well-being of the basin, yet each contributed to the degradation of the Willamette River, a process that we will examine in depth in the next section. However, several social characteristics of the basin combined to help save the river over the course of the 1900s. First of all, because almost all of the basin lies within the state of Oregon, the state can control what happens to the river. In many cases when water resources are polluted, a number of states and even countries must resolve differences before they can address the problem. Second, intense settlement did not begin along the Willamette River until the mid-1800s, much later than in many other areas of the United States. When the citizens of the basin began to be concerned about water pollution along with the rest of the United States, their river was not as irretrievably damaged as many others throughout the country.

LOOKING BACK: THE HISTORY OF THE WILLAMETTE RIVER BASIN

Although the Willamette did not fully recover from pollution and other mismanagement until the 1970s, its history is not only one of abuse, but also of public concern and efforts toward conservation.

Abusing the Resources of the Basin

Too often, uncontrolled use of a resource results in abuse, and such was the case with the Willamette. Logging and agricultural development entailed massive cuts of timber; the bare slopes that resulted were much more susceptible to soil erosion. As rain and melting snow washed more soil away from denuded hillsides and agricultural fields, the soil built up as sediment in the river. The excessive sedimentation contributed to the degradation of water quality through effects such as heightening BOD, smothering fish eggs, and preventing light from penetrating to rooted aquatic plants. In addition, chemicals often adhere to sediments and thus can increase the danger to aquatic habitats. Although sediments pose a minimal danger to human health because they can easily be removed by filtration, sediments dredged from some bodies of water can be so laced with dangerous chemicals they must be treated as hazardous wastes.

Agriculture also contributed to worsening water quality. Large-scale irrigation reduced low summer flows even further than normal, lowering DO and degrading water quality. In addition, dams blocked the passage of salmon migrating upriver to spawn, drastically reducing populations. However, by far the most serious threat to the waters of the Willamette was domestic and industrial waste.

Until 1939, domestic and industrial waste was commonly dumped untreated into the river or its tributaries. Only 52 percent of the total population were connected to a sewer system, and only 17 percent of these systems had any sort of water treatment facilities. Most of this waste, including wood fiber from the pulp and paper factories, substantially raised the BOD in the river. Because of its high concentration of fecal wastes, the polluted water was also a potential source of disease. Low summer flows exacerbated these problems. Poor water quality also helped to reduce fish and other aquatic populations.

Increasing Awareness of the Problem

By the late 1920s, widespread recognition of the river's degradation led citizens to form civic groups such as the Anti-Pollution League to explore and correct the problem. At the request of another civic group, the League of Municipalities, the Oregon Agricultural College (now Oregon State University) conducted a preliminary study of Oregon streams in 1929. Their report catalyzed a more in-depth study of water quality during the summer months of 1929.

Unusually thorough for the time, the resulting report confirmed that the Willamette was in trouble. Although DO content was high in the upper (upstream, or southern) river, two significant decreases in DO occurred just below (north of) Salem and Newberg due to waste from households and industries which increased the BOD. By the time the Willamette reached the Columbia River, DO in the water was less than 0.5 ppm, significantly less than the minimum 5.0 ppm needed to maintain healthy populations of fish and other aquatic life. The same study found that concentrations of coliform bacteria increased from 0 or 1 per milliliter to 100 per ml below small towns and 1000 per ml below the city of Salem. The presence of coliforms, bacteria that are normally present in high numbers in the large intestines of humans and other mammals, indicates the likelihood of contamination by infectious disease-causing organisms.

Despite these sobering findings, little was done to alleviate the problem until 1938, when the people of Oregon initiated the Water Purification and Prevention of Pollution Bill. This legislation set standards for pollution control and created the State Sanitary Authority to help take remedial action against pollution. Unfortunately, a lack of funds, unenforced regulations, and the advent of World War II all combined to prevent any real improvement in the state of the river. In 1944, a new study by the Engineering Experiment Station of Oregon State University showed that the river's condition had worsened since 1929. By mid-century, DO in Portland Harbor was found to be zero, and the upstream concentration at Salem was only 3.6 ppm.

Reviving the River

Eventually, the State of Oregon took measures to reduce pollution and improve flow which finally reversed the steady decline of water quality. Although the Sanitary Authority had required all municipalities to begin plans for sewage treatment facilities by the early 1940s, they were not all completed until 1957.

Meanwhile, the Sanitary Authority discovered that the pulp and paper industry was responsible for an inordinate share of the river's problems. The sulphite waste liquors that the industry dumped into the Willamette were responsible for about 84 percent of the oxygen demand on the river. In 1950, the Sanitary Authority ordered these factories to develop primary treatment facilities for those wastes and to limit discharges during the low-flow months of June through October.

Primary treatment is a physical process that removes undissolved solids such as sticks, rags, and other large objects. In some plants, the remaining sewage is chopped or ground into smaller particles; then dense material, like cinders, sand, and small stones, and undissolved suspended materials, including greases and oils, are removed. Undissolved organic materials sink to the bottom, where they form a sludge and are removed. The remaining liquid, called primary effluent, is often chlorinated before it is released. Primary treatment is about 50 percent effective at purifying wastewater; it cannot remove excess nutrients, dissolved organic material, or bacteria.

When further studies in 1959 showed no improvement in DO, BOD, or bacterial contamination, the Sanitary Authority ordered the cities of Eugene, Salem, and Newberg to install secondary sewage treatment facilities, and required Portland to accelerate completion of facilities already under construction. Secondary treatment is a biological process in which bacteria consume dissolved organic matter, and it is usually 85-90 percent effective. The Sanitary Authority also further reduced the amount of waste which pulp and paper operations could dump into the river. By 1967, Oregon required secondary treatment for any wastes dumped into the river or other public waters, and placed strict limits on the BOD that pulp and paper wastes could impose on the river. By 1972, all mills had installed secondary treatment equipment, which reduced BOD 80-94 percent from original levels. By 1980, 91 percent of all types of wastes were being removed from the discharge into the river.

Another way the Sanitary Authority worked to improve water quality was through regulating the rate of flow in the river. In the early 1960s, the state began to use reservoirs to regulate flow. The reservoirs were filled during the winter as part of a strategy to control flood waters. (In 1964, this strategy reduced a flood with a possible peak stage of 45.3 feet to 37.8 feet.) During the summer, controlled amounts of water were released from the reservoirs to the river, thus augmenting flow.

By augmenting summer flows and substantially reducing municipal and industrial wastes, Oregon raised water quality to acceptable standards at least 75 percent of the time between 1976 and 1980. This achievement was the result not only of well-enforced regulations, but also of the cooperation and concern of the industries in the Willamette Basin. For example, during a 1977 drought, major industries along the Willamette voluntarily reduced their waste output by 15 percent.

Graduating from Cure to Prevention

The 1970s were a time when the citizens of the basin were able to turn their attention from cleaning up old problems with the river to preventing new ones. Not only had water quality improved, but increased awareness helped solve other environmental problems as well. Dams were altered to include fish ladders, or series of steps along the sides of the dams. The steps allow water to flow down so that fish can jump from level to level, eventually reaching the top and continuing upriver to spawn. After the installation of the fish ladders, the fall chinook salmon run increased from a count of 79 in 1965 to more than 22,000 in 1973.

In addition, logging operations had begun to adopt sustained yield practices, which are methods of harvesting trees in which the number of trees cut does not exceed the capacity of the forest to regenerate itself. These practices helped to reduce erosion and sedimentation problems.

One program that helped maintain the quality of the Willamette River is the Willamette Greenway Program, a river park system established by the 1967 Greenway Statute. By ensuring the existence of vegetated areas along the river, the statute was designed to protect natural resources while allowing the public to enjoy and learn about them. In addition, the vegetation helps prevent erosion, provides habitats for wildlife, and enhances the beauty of the riverbanks.

Because the riverside areas are already heavily populated, the Greenway is a conservation corridor that is primarily privately owned, with areas of public parks and river access points. The Greenway Plan involves 510 riverbank miles, stretching from the Columbia River to the Dexter Dam on the Middle Fork and the Cottage Grove Dam on the Coast Fork. The plan safeguards all existing uses of the land, especially farmland.

During 1971, the Oregon Department of Transportation began acquiring land in the Greenway area for five major state parks. The state parks, three of which are located near heavily populated areas, encourage short trips for activities such as canoeing and wildlife watching.

In addition to the parks, each of the 19 cities and 19 counties involved has placed land available for public use into five categories, ranging from preservation (which includes sensitive areas protected as wildlife and vegetation preserves) to recreation (which includes areas that can be used intensively by the public). Between 1973 and 1979, 43 Greenway sites were constructed or improved; many of these provided opportunity for low intensity recreational activities such as hiking and river access through boat landings.

LOOKING AHEAD: THE FUTURE OF THE WILLAMETTE RIVER BASIN

The efforts of the people of Oregon to clean up the Willamette River have become a true success story. The Willamette is widely hailed as a positive example of what can be accomplished when the public, the government, and industry work together to solve environmental problems. But continued diligence is essential. Authorities must continue to monitor water quality. Although municipal and industrial wastes are under control, pollution can come from less obvious sources, such as fecal matter from grazing livestock or agricultural run-off containing fertilizers or nutrients. Another unexamined danger is heavy metals and other toxic materials that might accumulate in food chains.

Despite the importance of continued monitoring, the biggest challenge that the state of Oregon will have to meet is the basin's growing population, which is increasing at a faster rate than that of the nation as a whole. This growing population will produce more and more waste, thus increasing the potential for pollution. New ways to use or dispose of waste, such as using it for livestock feed, fertilizer, energy, or recycling, should be explored. An expanding population also threatens the existence of vegetated areas near the river. Besides serving important aesthetic and recreational purposes, these areas are necessary to prevent erosion. Although the Greenway program protects such areas, the public must remain interested in and educated about the Greenway so they will continue to support it.

Finally, the public must remain informed and interested in water quality. As history shows, the citizens were the driving force behind cleaning up the Willamette, and it is clear that their support will be necessary for continued monitoring and protection of the river. Government agencies and civic groups can help secure that support by continuing their efforts to educate and motivate the public about water resources.

SUGGESTED READING

Aikens, C. Melvin, ed. *Archeological Studies in the Willamette River Valley, Oregon.* Eugene, OR: Department of Anthropology, University of Oregon, 1975.

Bowen, William A. *The Willamette Valley: Migration and Settlement on the Oregon Frontier.* Seattle: University of Washington Press, 1978.

Brown, Carl, Joseph G. Monks, and James R. Park. *Decision-making in Water Resource Allocation.* Lexington, MA: Lexington Books, 1973.

Corning, Howard McKinley. *Willamette Landings: Ghost Towns of the River.* Second Edition. Portland, OR: Oregon Historical Society, 1973.

Experience the Willamette National Forest. Eugene, OR: The Forest, 1990.

Franklin, Karen E. "Fish Dammed by Dams." *American Forests.* August 1986. 11-12.

Gleeson, George W. *The Return of a River.* Corvallis, OR: Advisory Committee on Environmental Science and Technology and Water Resources Research Institute, Oregon State University, 1972.

Grove, Noel. "Greenways—Paths to the Future." *National Geographic.* June 1990. 77-99.

Palmer, Joel. *Journal of the Travels over the Rocky Mountains.* Ann Arbor, MI: University Microfilms, 1966.

Preliminary Willamette River Greenway. Oregon Department Of Transportation, 1974.

Rickert, David A. "Use of Dissolved Oxygen Modeling Results in the Management of River Quality." *Journal of the Water Pollution Control Federation.* January 1984. 94-101.

Upper Willamette River Basin. Salem, OR: State Water Resources Board, 1961.

Van Dyk, Jere. "Long Journey of the Pacific Salmon." *National Geographic.* July 1990. 3-37.

Willamette Basin Task Force. *Willamette Basin Comprehensive Study, Main Report.* 1969.

"Willamette River Greenway." *Oregon Lands.* June 1979.

CHAPTER 13

SOIL RESOURCES: MARATHON COUNTY, WISCONSIN

Soil is probably one of our least appreciated resources. Most of us notice it only when we track it into our homes, wash it from our clothes, or clean it off our automobiles. Nevertheless, soil is as essential for life as air or water; the food we eat, the clothes we wear, and the materials in our homes and automobiles all originate, directly or indirectly, in the soil. In addition, like air and water, soil can be degraded and depleted. In fact, although soil is normally a renewable resource, it can be abused to such an extent that it becomes, for all practical purposes, unrenewable. We degrade our soil when cover it with concrete and asphalt, when we poison it with pesticides, herbicides, and toxic wastes, when we let it wash or blow away, and when we allow it to become salty or waterlogged. Unfortunately, such activities are becoming more and more common as human populations grow and industrialization continues.

Although loss of soil through erosion, development, and other forms of degradation are worldwide problems, many techniques exist to combat and prevent them. In Marathon County, Wisconsin, local farmers who rely on the soil for their livelihoods have worked to protect it and retain it as a renewable resource. Located in north central Wisconsin, Marathon County is one of the state's biggest agricultural producers. Although Marathon farmers grow a wide variety of grains, they are best known for their vast output of dairy products. In fact, of all dairy-producing counties in the United States, Marathon County ranks fifth in total number of dairy cows and seventh in total milk production. Because of this heavy reliance on agriculture in general and dairy farming in particular, Marathon County residents realize that maintaining the fertility of their soil is vitally important.

Unfortunately, studies in the early 1970s revealed that at least some of Marathon County farms were losing soil much faster than it could be replaced. In accordance with Wisconsin 's long history of soil conservation and management, Marathon County established a number of programs aimed at preserving farmland, conserving soil, and reducing pollution caused by erosion. Such continued vigilance is an excellent example of what agricultural districts nationwide could do to preserve their precious farmlands.

PHYSICAL BOUNDARIES: HOW FERTILITY AND ERODIBILITY AFFECT MARATHON COUNTY SOIL

Marathon County occupies approximately 1 million acres in the heart of Wisconsin. The county owes its agricultural success in part to a climate with long, relatively cold winters, warm summers, and steady, ample rainfall. However, the rich soil remains the most precious resource for Marathon county farmers.

Soil Characteristics

The activity of a series of glaciers is one factor responsible for the character of Marathon County's soil. About one million years ago, a massive glacier covered and retreated from most of the midwest, including Marathon County. However, it was subsequent glaciers that shaped Marathon County as it exists today. The glacier that occurred during the period known as the Illinoian stage covered all of Marathon County, flattening hills and depositing drift that became the basis for today's soil. The western part of the county, untouched after this process, is a flat to gently rolling area with rich soils that make it the major agricultural area of the county. However, the glacier of the subsequent Wisconsin stage did affect the eastern part of the county, leaving behind a narrow ridge of transported soil and rocks called a moraine. To the east of this moraine lies a pattern of swamps, hills, and scattered "pothole" lakes. When this glacier melted, about 15,000 years ago, the melting water formed what we now call the Wisconsin River.

Soil in Marathon County has also been created by the weathering of the Canadian Shield, a huge sheet of igneous and metamorphic bedrock which underlies most of Canada and the northern United States. In places where the bedrock was exposed, wind and rain wore off small particles that served as the mineral base for new soil to form. Additional soil has been carried into the area by water and rain.

The major type of soil found in Marathon County is known as alfisol, which typically has a good moisture supply and is relatively fertile. Because this type of soil is also moderately acidic (with a pH range of 4.5-6.5), farmers must lime their fields heavily to neutralize the acidity. The texture of Marathon County soil is generally loamy, with silty loams found in the western part and sandy loams in the east. Loamy soils are comprised of about 40 percent sand (sand particles range in diameter from 0.05 mm to 2 mm), 40 percent silt (silt particles range from 0.002 mm to 0.05 mm in diameter), and 20 percent clay (clay particles are smaller than 0.002 mm in diameter). Loams have an excellent texture for growing most crops because they provide adequate air spaces and allow good drainage while also retaining enough moisture for plant growth. (In contrast, soils with higher clay contents may retain too much water and become compacted, whereas soils with higher sand contents are typically too porous to retain sufficient moisture.)

Although texture and porosity help determine soil fertility, the upper, most fertile layer (topsoil) must also retain sufficient levels of its three most important components: humus, living organisms, and some minerals. Humus consists of decayed and decaying organic matter, including dead plants, animals, and other organisms. Because it helps to retain water and to maintain a high nutrient content, humus is an important part of fertile soil. Humus content in Marathon County ranges from 3-4 percent in the western part of the county to 2 percent or less in the east.

Living organisms are also essential to maintaining healthy soil. Most organisms found in soil are microorganisms such as bacteria and fungi that decompose organic material, thus recycling the nutrients in this material and producing humus. In addition, earthworms turn over the soil, mixing fertile topsoil with deeper, less fertile soils and helping to aerate the soil. The digging and burrowing of small mammals such as moles, insects and their larvae, and other arthropods also aerate and drain the soil. Finally, the wastes of all these organisms contain nutrients that are recycled back into the soil, as are the organisms' bodies after they die.

Although much of Marathon County's soil is fertile and otherwise suitable for growing crops, approximately 25 percent is unsuitable because of factors such as lack of fertility, inappropriate terrain (too sloped or rocky, for instance), or a tendency to erode. However, much of this land can be put to less intensive uses, such as grazing herds of cattle.

Soil Erosion

Soil erosion, an issue that deeply concerns residents of Marathon County, is also a major problem in the United States and throughout the world. When soil is not held down by roots of grass, trees, or other plants, it tends to wash or blow away (especially on hills or mountain slopes). Although all soil experiences some erosion, undisturbed areas rarely experience net losses because natural processes such as weathering and decomposition usually replace soil as fast as or faster than it vanishes. The amount of soil that an area can lose through erosion without a subsequent decline in fertility is known as the soil loss tolerance level (T-value) or replacement level. Depending on the type of soil, an acre of land may have a T-value between 2 and 5 tons/acre/year, with natural processes compensating for that loss with the production of new topsoil. Although natural causes such as severe droughts can lead to soil erosion above replacement level, human activities are more often responsible for accelerating soil loss above T-value.

Agriculture is a prime culprit in human-caused soil erosion, although development, deforestation, and other human activities play a part as well. One major cause of agricultural soil erosion is the cultivation of marginal or poor cropland. Farmers worldwide have attempted to maintain or increase production by cultivating steep slopes, rain forests, wetlands, and other areas that are not suited to agriculture. In addition, even good cropland experiences erosion if it is farmed in inappropriate ways. Poor farming techniques, such as allowing fields to lie bare
and exposed to the wind, rain, and snow in winter, are all too common in modern agriculture.

Soil experts estimated that the arable, or fertile, land in the United States had lost one-third of its topsoil layer between 1776 and 1976. Currently, erosion from poorly protected lands accounts for the loss of approximately 3 billions tons of topsoil per year; although this rate of loss is actually greater than that experienced during the Dust Bowl of the 1930s, it is less noticeable because today more soil is lost by water erosion than by the more visible action of wind. Of the approximately 400 million acres of productive farm land in the United States, about 100 million acres are eroding at twice the rate at which they can be replaced by new soil formation, and another 90 million acres are eroding at one to two times the replacement level. Some croplands on highly erodible land are being lost at four times the replacement level. Such soil erosion can have a dramatic effect on fertility and productivity. According to the Worldwatch Institute, a Washington-based nonprofit research organization, for every 1 inch of topsoil lost, yields of wheat and corn drop by about 6 percent.

The soil in Wisconsin is no exception to the widespread threat of soil erosion. Croplands are the major problem, responsible for as much as 85 percent of the eroded soil in the state. Studies link an increase in erosion with an increase in the amount of land planted to row crops, especially corn and soybeans; the percentage of land planted with these types of crops grew from 31% in 1970 to 46% in 1982. Especially when rows run up and down hills, they encourage water erosion.

An estimated 64 million tons of soil are lost from Wisconsin cropland annually. Although wind erosion carries away about one-fourth of this soil, the remaining 48 million tons are lost to water erosion. Extensive water erosion occurs on cropland almost exclusively because of the lack of plant cover and root systems; after the crop is harvested, many farmers plow the fields, leaving bare soil. Root systems not only hold soil in place, but also pull down water so it washes away less soil. A heavy rain can wash away precious topsoil and even newly planted seeds or seedlings that haven't put down firm roots. Water erodes soil in several ways. Sheet erosion occurs when falling raindrops dislodge soil particles. When rain falls faster than it can be absorbed by soil, sheets of water sweep away the dislodged soil. Heavy sheets or runoff can cut grooves or rills into the soil. As the runoff carries away more and more soil, the rills may form large gullies.

Soil losses in Wisconsin may be higher than the national average because much of the cropland is on sloping land. At current erosion rates (including both wind and water), Wisconsin can expect to experience significant soil fertility losses in 20 to 50 years. The highest erosion rates occur in the southern part of the state, where erosion averages between 8.1 to 9.3 tons per acre per year. At the other extreme, erosion in many of the northern counties averages between 0 and 3 tons per acre. These figures are only averages, however, and don't always show the whole picture. Actually, most soil erosion occurs on a relatively small percentage of agricultural land. About 2.2 million of the state's 11.5 million acres of cropland are eroding at over twice the replacement level; of Wisconsin's 64 million tons of eroded soil, 40 million are lost from these areas. In other words, 60 percent of the erosion occurs on less than 20 percent of the land! To achieve a significant drop in overall rates, control measures must be focused on areas with the highest erosion rates.

In Marathon County, actual erosion rates range from 1 to 16 tons per acre per year, depending on the the topography and crops grown, with an average soil loss of 2.7 tons per acre. Replacement level, or T-value, in Marathon County averages 3 to 4 tons per acre. Approximately 16,900 acres, or 5 percent of the county's total cropland, are eroding at rates greater than the T-value.

BIOLOGICAL BOUNDARIES: NATURAL RESOURCES
THAT PROTECT AND DEPEND ON THE SOIL

Farmland and forests account for 9 out of every 10 acres in Marathon county, allowing a wide variety of plants and animals to flourish in the area. These plants and animals interact with the soil in many ways, both contributing to fertility and depending upon it.
The forested areas principally contain hardwoods and evergreens that have grown back after

an original cutting, sometimes hundreds of years ago. The county's 197 lakes and 256 rivers and streams are populated by hundreds of species of fish, including such popular sportfish as trout, bass, pike, and panfish. These waters are also home to many types of waterfowl and attract flocks of migratory geese. Wild animals, ranging from chipmunks and rabbits to white-tailed deer and black bears, can also be found in the county.

Farms occupy over 50 percent of the land in Marathon County. Although many of these farms yield large crops of oats, corn, hay, barley, potatoes, and wheat, the grain is primarily used to feed dairy cows. Some farm acreage is also used to graze herds of cattle. The dairy industry produces approximately 65 percent of the county's agricultural cash income.

In addition to its reputation as a major dairy producer, Marathon County acts as one of the major producers of ginseng in the United States. Asian ginseng, a perennial herb, has been prized for its medicinal properties in the Orient for thousands of years. American ginseng, a related variety, is native to the northeastern United States, including Wisconsin. Ginseng is generally thought to increase the body's ability to deal with stress and fight off disease. Because ginseng takes at least 5 years to produce a marketable root and must be grown in highly shaded areas, cultivating this crop is labor and capital intensive. Although ginseng has been declared an endangered species and special permission must be obtained to sell it outside the country, 95 percent of the ginseng produced in the Unite States is exported, primarily to the Orient. The approximately 2000 growers in Wisconsin produce 95 percent of the cultivated ginseng in the United States, and 1200 of these growers are in Marathon County.

SOCIAL BOUNDARIES: THE INTERACTION OF AGRICULTURE AND DEVELOPMENT

Although Marathon County is primarily an agricultural district, somewhere between 125 and 150 thousand acres are reserved as open space and recreation areas. Only about 32,000 acres—3.2 percent of the county's land—is used for nonfarm development. Roughly half of the 112,000 people who live in the county live in the Wausau urban area, which includes the cities of Wausau, Schofield, and Rothschild, and the townships of Weston, Rib Mountain, and Stettin.

Despite the rural nature of the county, a trend toward nonfarm uses that began in the 1960s poses a threat to its prime farmland. Land that once produced 150 bushels of corn per acre has been converted to urban development, roads, and housing. In 1982, Marathon County had 3300 working farms. By 1987, the number of working farms dropped to 3000 because of a combination of economic factors, urban development, and consolidations. Although the size of the average farm has been increasing as the number of farms drops, there has been an overall decrease in land used for agriculture; between 1961 and 1981, farm acreage in Wisconsin declined by about 3.5 million acres (15 percent). Although corporate farms and farm partnerships have increased, over 90 percent of Marathon County's farms are still family-owned, with less than 10 percent tenancy or absentee ownership.

LOOKING BACK: THE HISTORY OF MARATHON COUNTY AND ITS SOIL

Marathon County is a microcosm of the problems facing agricultural land throughout the United States: soil erosion, runoff laden with fertilizer and animal waste, encroachment by urban development, and the trend toward specialization and conglomeration. However,

Marathon County's history and extensive network of farmland preservation efforts can also serve as a model for effectively coping with these problems.

A Tradition of Soil Conservation

In the 1850s the first wave of German and Polish farmers arrived in Marathon County, anxious to create a pastoral landscape similar to those they had left behind. The fact that cleared land was considered of less value for tax purposes also hastened the transition of the county from a forest products-based economy to a farm products-based economy. Accordingly, agriculture has been the dominant influence on Marathon County since the late 1800s.

Awareness of the need to preserve soil quality in Wisconsin also began in the 1800s. As early as 1880, German farmers in LaCrosse County planted crops in strips across the slopes of hills in order to control erosion. In 1890, F.H. King of the University of Wisconsin began erosion and soil conservation studies. Then, in the early 1920s, O.R. Zeasman made soil conservation an important issue in the state; as a result of his influence, gully control structures and terraces to prevent soil loss were built throughout Wisconsin.

Building on the state's tradition of soil conservation, in 1931 the Wisconsin Agricultural Experiment Station joined with the U.S. Department of Agriculture to study erosion and ways to control it, establishing a Soil Conservation Experiment Station on Grand Dad's Bluff at LaCrosse. In 1933, nine Civilian Conservation Corps camps were set up in Wisconsin to fight erosion, and the Soil Conservation Service (SCS) began the nation's first large-scale demonstration of soil and water conservation on 90,000 acres in Coon Valley. Coon Valley had lost approximately 3 inches of soil from fields that had been farmed for 80 years. Persuaded by low yields and incomes, area farmers agreed to work with the SCS to develop and apply whole-farm conservation plans.

The conservation plans they used included a number of methods still recommended today. The most effective way to prevent soil loss, of course, is simply not to farm marginal croplands, especially slopes and soils with structures not suited to cultivation. Another effective measure is to plant a cover crop—a crop that covers the ground during the winter after the main crop is harvested. The roots of the cover crops anchor the soil, protecting it from the effects of rain, snow, and wind. In the spring, the crop litter and crop residues can be kept on the ground surface, instead of being tilled under, to act as a green manure. Planting trees and shrubs along the windward side of fields works well against wind erosion. The most effective windbreaks, or shelterbelts as they are sometimes called, help to deflect the wind upward while slowing its velocity.

Several alternative methods of tilling can help to prevent soil erosion. Ridge tilling involves planting the crop on top of ridges that run perpendicular to the direction in which rain and snow run-off flows. Conservation tillage includes both low-till and no-till methods of planting. Low-till planting consists of tilling the soil just once in the fall or spring, leaving 50 percent or more of previous crop residue on the ground's surface. No-till planting, or stubble mulch farming, is done with a no-till planter or drill that can sow seeds without turning over the soil. Crops are planted amid the stubble of the previous year's crop, which acts as a mulch to fertilize the soil and prevent it from drying out. Unfortunately, because the previous year's mulch also provides food and cover for pests, insects, and weeds, conservation tillage typically requires the intensive use of pesticides or herbicides.

Not surprisingly, the risk of erosion is greatest on land that is sloped. When sloped land is brought under cultivation, a number of techniques can be used to reduce that risk. In stripcropping, rows of grain are alternated with low-growing leaf crops or sod, which offer greater protection to the soil. These strips of grasses or legumes also help to improve soil structure and and enhance the organic content of the soil. Stripcropping can be used in conjunction with contour plowing or terracing in order to offer increased protection from erosion. Contour plowing involves tilling the soil parallel to the natural contours of the land rather than in the straight rows and squares characteristic of conventional fields. Contour plowing helps to keep the soil from washing down the hillside. When the slope is extreme, contour terracing can be used. Contour terraces are broad "steps" or level plateaus built into the hillside. Swales, or trenches, located at the edge of the terraces act as catch basins for the rain, channeling it along the hillside. Swales planted with grass, called grassed waterways, help to both slow the course of the water and absorb some of the excess water and any eroding soil.

As soil conservation projects using some of the above techniques sprang up throughout the state, farmers learned the importance of working with neighbors and communities to control erosion, a problem that is rarely confined to just one farm. Accordingly, in 1937, the Wisconsin Legislature enacted a law providing for the organization of each county into a soil conservation district to help county residents work effectively against erosion. All districts were to have working agreements with SCS and other groups that could give technical, financial, and educational assistance. The Marathon County Soil and Water Conservation District (SWCD) was established in 1941.

Protecting the Soil Today

The amount of agricultural land in Marathon county expanded steadily until the early 1950s; even land only marginally suited for agriculture was cultivated. By the end of the 1950s, advances in machinery, seed quality, fertilizers, and pesticides made farming more profitable and less physically demanding. Marginal lands were once again taken out of production.

In the early 1970s, people throughout the state became aware that increasing nonfarm development was threatening to remove prime farmland from the agricultural base. In addition, some farmland was experiencing rates of erosion much higher than the T-level. To eliminate these threats before they could damage the state's precious farmland, the State of Wisconsin passed the Farmland Preservation Act in 1977. This act created a program of tax incentives for landowners who agree to preserve agricultural land.

Farmers can become eligible for tax credits in two ways: if their land is zoned for exclusive agricultural use, or if they sign a contract agreeing not to develop their land for a specific amount of time. To qualify for a contract, a farmer must have at least 35 acres that have produced at least $6000 worth of farm products in the last year or $18,000 worth in last three years; in addition, all participants must conform to soil and water conservation standards adopted by the county.

Marathon County completed its own Farmland Preservation Plan in 1982. Goals call for (1) the preservation of prime agricultural land with soil productivity as the basis of preservation, (2) support of family farm ownership, (3) reduction of the erosion of topsoil, and (4) the wise use of farmland in urban fringe areas.

A comprehensive map of the county was prepared to delineate farmland preservation areas, urban growth areas, and special environmental areas (wildlife habitat, wetlands, scenic and historic sites). Implementation is optional, but as of 1987, if farmers participate in the program, they must adopt a comprehensive soil management plan that reduces soil loss to T-values before tax incentives are granted. From 1982 to 1987 over 600 Marathon County farms took advantage of the state's Farm Preservation Program, which is recognized as one of the most effective and innovative in the United States.

Marathon County is also participating in the "T-by 2000" erosion control program passed by the state legislature in 1982. By the year 2000, soil erosion rates on Wisconsin cropland are to be reduced to the soil-loss tolerance level. Practices that can reduce erosion rates include conservation tillage, strip cropping, water runoff diversions, terraces, windbreaks, and permanent vegetative cover. Use of these practices can reduce erosion to 10 percent or less of pre-practice erosion rates.

Approximately 1000 of the farms in Marathon County have met conservation compliance requirements because of the 1985 Food Security Act. This act applies only to farmers participating in federal farm commodity programs, programs that ensure that farmers receive a minimum price for crops. The act requires farmers with land identified as highly susceptible to erosion to develop and implement conservation plans for the highly erodible fields. The conservation plans typically include a variety of measures, such as conservation tillage, stripcropping, terracing, and grassed waterways. Another effort, the Conservation Reserve Program (CRP), offers yearly payments to farmers who voluntarily take highly erodible land out of production. The roughly 2000 farms not included in federal programs also contain some highly erodible land, but these farmers can choose whether or not they will make an effort to preserve soil.

In the 1980s, Marathon County residents also became concerned about the pollution of streams, lakes, and groundwater by rain and snow runoff containing nutrients and animal wastes. Animal wastes and nutrients such as chemical fertilizers containing nitrogen can stimulate massive algal growths in bodies of water. Such growths deplete the oxygen in the water, often killing off other life forms and upsetting the balance of the ecosystem. The county's 88,000 dairy cows were of particular concern because of the sheer amount of waste they produced: approximately 4000 tons of manure per day. In response to the concern about water pollution, the county adopted an animal waste ordinance, which required permits for the construction of animal waste storage facilities. Such permits ensured that wastes were properly stored and utilized.

The Marathon County Animal Waste Plan, another program aimed at reducing water pollution from animal waste, acquired funding from the Wisconsin Department of Agricultural Trade and Consumer Protection. This plan evaluated watersheds in the county for their susceptibility to problems due to animal waste as well as their potential for improvement. The Little Rib River Watershed was selected and landowners in that area are eligible for cost-sharing for storage facilities and barnyard improvement projects.

The county is also administering a program initiated by the Wisconsin Department of Natural Resources to protect watersheds vulnerable to pollution, especially pollution from agricultural runoff and erosion. Watersheds selected through 1991 include the Upper Big Eau Pleine Watershed, Lower Big Eau Pleine Watershed, and the Upper Yellow River Watershed.

In each watershed, all pollution sources are inventoried, as well as areas such as wetlands that have a positive impact on water quality. After all factors that affect water quality are evaluated, a list of landowners eligible to participate in the project is prepared. Upon signing contracts, these landowners receive up to 70 percent state cost sharing for instituting conservation practices.

Faced with such a variety of programs designed to conserve soil and prevent runoff from polluting water, Marathon County established a Soil Erosion Control plan in 1988 to coordinate activities. The plan's purpose is to coordinate and implement the federal, state, and county programs while also continuing to identify land where erosion is greater than replacement levels.

In 1991 the county started a unique project to reduce the total amount of nitrogen entering groundwater, lakes, rivers, and streams. A full time staff member was assigned to work with landowners that accept wastes generated outside their farms. These wastes typically include sewage sludge from treatment plants, cheese factory waste water, whey permeates, and slaughter plant wastes. Then comprehensive plans are developed to account for and limit all the sources of nitrogen on a particular farm, including off-farm wastes, animal manure, and fertilizers.

LOOKING AHEAD: THE FUTURE OF SOIL CONSERVATION IN MARATHON COUNTY

Marathon County certainly does not have the worst erosion problems in the country. Counties in California, Iowa, Washington, and Tennessee have experienced greater soil loss, both in total tons of topsoil and in acres of prime farmland. However, Marathon County land holds the potential for severe soil fertility loss because its topsoil is not naturally deep, averaging just 8 inches. Any sustained loss of topsoil will have serious effects on soil fertility and productivity and thus have a devastating impact on the county's economy.

Consistent and timely concern by both the state and the county have kept this sort of erosion from taking place so far. Through its various programs, Marathon County provides an example of the kind of soil management and farmland preservation efforts needed to combat erosion, maintain soil productivity, preserve family farms, and protect waterways, wetlands, and historic places. Of course, erosion will always be a threat, so continued vigilance is essential even after soil loss has been reduced to replacement levels throughout the county.

SUGGESTED READING

Marathon County Land Conservation Committee. June 1988. *Marathon County Soil Erosion Control Plan.*

Marathon County Land Conservation Department. *Marathon County: Animal Waste Management Plan.* June 1987.

Marathon County Planning Commission. *Marathon County Farmland Preservation Program.* 1982.

Marathon County Soil and Water Conservation District. *Resource Conservation Program: A Long Range Program for the Resources of Marathon County.* 1975.

Morgan, R.C.P. *Soil Erosion and Conservation.* Ed. D.A. Davidson. New York: Wiley, 1986.

——, ed. *Soil Erosion and its Control.* New York: Van Nostrand Reinhold, 1986.

Nesbit, Robert C. *Wisconsin: A History.* Second Edition. Revised and updated by William F. Thompson. Madison, WI: University of Wisconsin Press, 1989.

Reganold, John P., Robert I. Papendick, and James F. Parr. "Sustainable Agriculture." *Scientific American.* June 1990. 112+.

Sattaur, Omar. "Erosion and Yield, a Tricky Relationship." *New Scientist.* June 3 1989. 47.

Sinclair, Ward. "Keeping Soil Down on the Farm." *Sierra.* May/June 1987. 26-29.

Soil Conservation Service. *Soil and Water Conservation in Wisconsin.* Madison, Wisconsin: February 1965.

"Soil Erosion Could Be Eliminated by 1995." *Earth Science.* Summer 1990. 8-9.

Trimble, Stanley W., and Steven W. Lund. *Soil Conservation and the Reduction of Erosion and Sedimentation in the Coon Creek Basin, Wisconsin.* Alexandria, VA: U.S. Department of the Interior, Geological Survey, 1982.

Wisconsin Department of Agriculture, Trade, and Consumer Protection, Land Resources Bureau. *Soil Erosion Control Program.* September 1983.

Wisconsin Land Conservation Board. *Soil Erosion in Wisconsin.* October 1 1984.

CHAPTER *14*

MINERAL RESOURCES: CRITICAL AND STRATEGIC MINERALS

When most of us think of our environment, we tend to visualize its living components: the trees that make up a forest, the fish that inhabit a lake, or the grasses waving on the prairie. However, nonliving components play an essential role in our environment as well. Many of these nonliving components of our biosphere can be classified as minerals. In the broadest sense, mineral resources include all nonliving, naturally occurring substances that are used by humans. This definition covers a wide spectrum of substances, from basic materials such as granite and sand to precious metals such as gold and silver to fuel resources like oil and coal derived from once living matter.

Most of us are very familiar with fuel minerals such as oil, natural gas, and coal. Nonfuel minerals can be either metallic, such as copper, bronze, aluminum, lead, and iron, or nonmetallic. Nonmetallic minerals include a wide variety of resources. Sodium chloride, phosphates, nitrates, and sulfur are used for chemical, fertilizer, and special uses. Cement, sand, gravel, gypsum, and asbestos are building materials.

Minerals can be found everywhere—the oceans, the highest mountain peaks, and the air—but they are typically found bound up in rock within the earth's crust. Areas of rock that contain minerals are known as ores. Although the image of minerals caught up in layers of rock may seem unchanging and permanent, even the composition of rock slowly changes over the ages. Sediments are continually being eroded from the earth's continents by the atmospheric forces of water, wind, and ice. Most sediments eventually come to rest on the ocean bottoms along the continental margins, and are compacted and cemented to form sedimentary rock. As more sediment accumulates, increasing pressure and rising temperatures produce physical and chemical changes in the underlying sedimentary rock. Material near the bottom may even melt to form magma, which tends to rise out of the sedimentary rock because its liquid form makes it less dense than the rock. As the rising magma cools, it forms a new type of rock called igneous. The igneous rock eventually erodes to produce sediments, completing a cycle that takes place over millions of years.

Although minerals cycle and change with the rock layers of the earth over the centuries, many metals and nonmetallic materials also move through the food and energy cycles of different ecosystems. These materials, such as iron and nitrogen, are absorbed by living species through the food chain and are then returned to the ecosystem through waste or decomposition.

In addition to their integral role in the earth's cycles, many minerals have become essential to the functioning of human society. Not surprisingly, industrial countries consume the greatest share of mineral resources. It is estimated that the United States, with just 5 percent of the world's population, uses about 30 percent of the global production of minerals annually.

Many minerals and their uses are well known: iron and coal are used in steel production; aluminum is used in such diverse products as cans and automotive parts; and nitrogen and phosphorous are ingredients in chemical fertilizers. But many other, less familiar minerals are essential to industrial societies. Titanium, manganese, cobalt, magnesium, platinum, and chromium are vital to industrial processes and as components in aircraft, automobile engines, and other high-tech applications. Limestone, gravel, sand, and crushed rock, commonly used in the construction industry, are also important commodities for the building and maintenance of roads, highways, and bridges.

Industry relies heavily on approximately 80 minerals. Three-quarters of these either exist in relatively abundant supply to meet all our anticipated needs, or can be replaced by existing substitutes. In contrast, critical minerals are those that are essential to a nation's economic activity but which exist in relatively short supply; strategic minerals are those that are essential to national defense but which exist in relatively short supply. Most critical or strategic minerals are located in politically volatile regions, a factor making their continued supply undependable. Another factor that makes certain kinds of minerals essential is a lack of suitable substitutes. In this chapter, we will focus on four minerals—cobalt, chromium, manganese, and platinum group metals—that are widely considered to be the most critical or strategic to the United States today. In addition to highlighting the importance of mineral resources to industrial societies and military security, our exploration of the role of these four minerals also addresses the political controversies that can ensue over where and how mineral resources are extracted from the earth.

PHYSICAL BOUNDARIES: U.S. DEMAND AND SOURCES OF STRATEGIC MINERALS

Although the United States recycles varying proportions of chromium, cobalt, manganese, and the platinum group metals, it is almost completely import dependent for new supplies of these metals. In addition, world production of these metals is dominated by just a few countries, including the former USSR, South Africa, Zaire, and a few other nations in southern African .

Chromium, a bluish-white metal, is not found naturally in its pure form but rather as chromite ore. Chromite must be processed into ferrochromium before it can be used by industry. As of 1988, the United States imported over 75 percent of the chromite and ferrochromium it used; over 50 percent was imported from South Africa, with significant amounts also coming from Zimbabwe and the former USSR. The remaining percentage of chromium used in the United States was obtained from recycling.

Cobalt is a silver-white metal with exceptional properties of strength and hardness. Rarely found by itself, cobalt is usually mined as a by-product of copper or nickel sulfides. As of

1984, the United States produced no cobalt, importing 84 percent of its supply and obtaining the rest from recycling. Zaire, which possesses the world's greatest reserves of cobalt, supplies 37 percent of U.S. imports. Nearby Zambia supplies 12 percent.

Manganese, a silver-white metal, is commonly found in the form of oxides, such as manganese dioxide. The United States imported a full 100 percent of its manganese supply in 1988, obtaining most of that supply from South Africa and Gabon (also in southern Africa). Statistics show that U.S. reliance on South Africa, the major world supplier of manganese, is growing.

Platinum group metals, often known as the "noble" metals, have similar properties and often occur together in nature. They are platinum, iridium, and osmium, palladium, rhodium, and ruthenium. They are usually by-products of nickel mining. In 1988, the United States imported 93 percent of its platinum group metals, purchasing over 50 percent from the leading world producer, South Africa, and approximately 13 percent from the former USSR.

BIOLOGICAL BOUNDARIES: THE EFFECTS OF MINING DOMESTIC DEPOSITS OF CRITICAL AND STRATEGIC MINERALS

Although the United States produces almost none of the chromium, cobalt, manganese, or platinum group metals it uses, low grade deposits of all of these metals except manganese do exist domestically. Low-grade deposits are ores that contain a relatively low amount of minerals per given volume. These low-grade domestic deposits of cobalt, chromium, and platinum-group metals are generally not mined because it is more economical to buy them from abroad. In addition, many deposits lie on federally owned lands, where mining would be a source of controversy. Environmentalists fear that exploration and mining these deposits would have a variety of detrimental effects on the environment.

Many of these lands contain endangered species, like the grizzly bear found in some National Forests and wilderness areas in the west. The roads, traffic, and the mining operations themselves might disrupt animals' lifestyles, hampering reproduction or even endangering the animals' lives. Also, wild animals that become overexposed to humans and human activities lose their natural fear, becoming potentially dangerous to humans and more likely to fall prey to hunters.

Mining can also disrupt the animals' habitat. Even though most minerals are mined underground, the mining machinery pollutes the air, and run-off from debris and ore can seriously degrade water sources. Mining operations can add excessive noise to the nearby area and usually destroy some trees and other vegetation.

SOCIAL BOUNDARIES: WHY CRITICAL AND STRATEGIC MINERALS ARE ESSENTIAL TO THE U.S. ECONOMY AND NATIONAL DEFENSE

Although not every use of chromium, cobalt, manganese, and platinum group metals is critical, these metals are all essential to the U.S. economy and national defense. The fact that our chief suppliers have histories of political instability or poor relations with the United States makes our reliance on these metals a matter of even greater concern.

How We Use Critical and Strategic Minerals

Chromium, when mixed with other metals, gives them the sought-after qualities of hardness and resistance to high temperatures, wear, and oxidation. Chromium is an essential ingredient in stainless steel, forming a chromium oxide film on the surface which provides protection against corrosion and oxidation. Combined with nickel, aluminum, cobalt, or titanium, chromium forms superalloys that can withstand especially high temperatures. These alloys are used for a variety of products, including jet engine casings and turbine blades, ball bearings, and high-speed drills. Chromium is also a component of exhaust systems and some dyes and pigments, and acts as an agent in chemical processing, gas and oil production, and power generation.

Cobalt, like chromium, is known for its strength and resistance to corrosion. It is an essential component of jet engines; a Pratt and Whitney F100 Turbofan aircraft engine uses 885 pounds of cobalt. It is also used in nuclear control rods, computers, magnets, and many types of electrical equipment and acts as a catalyst in petroleum refining and other chemical processes. Cobalt alloys are used to make high speed tools and are popular for medical and dental work because the body rarely rejects implants made of cobalt. Cobalt is also added to glass, ceramics, livestock food, and fertilizers.

Manganese is essential for producing steel; in fact, 90 percent of the manganese consumed in the United States is used for the alloying and processing of steel. Manganese helps prevent the formation of iron sulfides, which weaken steel, by combining with the sulfides. It also removes oxygen and helps harden the final product. Because it becomes harder with pounding, manganese is also an important part of railroad equipment, rock crusher parts, and the teeth of power shovels. As manganese dioxide, it is used in dry cell batteries, photo development, and the production of rubber and plastic. Forms of manganese are also added to livestock feed and fertilizers.

Platinum is probably best known for its use in jewelry, but platinum group metals are also an important part of catalytic converters, which control auto emissions. Platinum group metals also serve as catalysts in chemical and petroleum refining processes. Platinum is frequently a component in equipment for the electronics and communications industries. Because platinum group metals are so expensive, they are generally used only when there are no satisfactory substitutes.

How Political Forces Affect the U.S. Supply of Critical and Strategic Minerals

Obviously, a cut-off in our supply of these important minerals would effectively halt the production of many products essential to our society: automobiles, airplanes, steel, and petroleum, among others. However, these minerals are found chiefly in countries that the United States cannot necessarily rely on for an uninterrupted supply.

Beginning with the end of World War I, the former USSR and the United States considered each other competitors in a "cold war" for superpower status. Before the Berlin Crisis of 1949, when the two superpowers first confronted each other over control of West Berlin, the United States imported 31 percent of its manganese, 47 percent of its chromium, and 51 percent of its platinum from the USSR. Following the conflict, the Soviets cut off this supply, but the United States simply turned to small producers like India and Turkey, which expanded production to meet their new market. Today, many small producers no longer have the resources to make up for the loss of a major supplier. Although the improvement in relations between the United States and the USSR throughout the 1980s helped to allay fears of cut-offs, the recent breakup of the USSR leaves the future uncertain. Each separate state that produces minerals important to us will forge its own new relationship with our country.

Another change that affected our critical or strategic mineral supply in the past was the post-World War II move towards independence of many nations that were formerly colonies of other countries. By the 1970s, over 100 new nations had emerged. Many of these new nations exported raw materials, including strategic ones, to the United States. Unfortunately, the old colonial system was not always replaced by stable governments. Many nations were split into factions and plagued by civil unrest, disorganization, and poverty. For instance, in 1978, anti-government guerillas invaded Zaire's Shaba province, the principal location of cobalt production, preventing any cobalt from leaving the country. As a result, the price of cobalt rose from $6 to over $45 per pound.

While the temporary disruption in cobalt from Zaire was due to unrest, a different sort of political problem occurred when Rhodesia announced its independence from Great Britain in 1966. Because Rhodesia vowed to continue white minority rule, the United Nations passed a resolution requiring all members to refrain from trade with Rhodesia. As a result, the United States could not buy Rhodesian chromium until the ban lifted in 1971. Fortunately, the government did not need large amounts of chromium at that time and managed to maintain its supply by buying from Turkey and the Philippines. Also, the steel industry developed a new method of producing stainless steel which used high carbon ferrochromium rather than the more expensive low carbon variety normally purchased from Rhodesia.

South Africa, our single greatest source of chromium, manganese, and platinum group metals, has always been a steady supplier. However, like Rhodesia in 1966, South Africa maintains a racist government through apartheid, a class system under which the ruling white minority legally discriminates against the black majority. In 1985, South Africa threatened to stop mineral sales to any country imposing economic sanctions due to apartheid. Because the United States as a nation has never imposed economic sanctions on South Africa, our supply of strategic minerals continues to flow from that source. However, increased internal resistance to the apartheid system, combined with growing global disapproval of any trade with South Africa, may well threaten our reliance on that country for strategic minerals in the future.

LOOKING BACK: THE HISTORY OF CRITICAL AND STRATEGIC MINERALS IN THE UNITED STATES

When the Industrial Revolution took hold in the late nineteenth century, the United States was an enthusiastic participant, building factories and developing technology at an ever-increasing rate. One reason the United States was so successful in industrialization was its abundant supply of natural resources to use as inputs for manufacturing. However, World War I made the United States realize that it depended on some minerals that could not be produced domestically in significant amounts. Since that time, opinions on the strategic mineral situation have fluctuated between the extremes of panic and complacency.

How the United States Developed a Stockpile

In 1921, the U.S. War Department drew up the Harbord List, a list of 28 minerals that had been in short supply during the war. In 1939, the government gave the navy $3.8 million to purchase reserves of important materials like tin, ferromanganese, tungsten, chromite, optical glass, and manila fiber. Then, in 1939, Congress passed the first stockpiling act, granting $70 million in order to establish a back-up supply of strategic materials including chromium and manganese. To further strengthen stockpiling efforts, President Franklin Roosevelt ordered the Reconstruction Finance Corporation in 1940 to conduct large-scale purchases of materials deemed necessary for war. One branch, the Metals Reserve Corporation, was specifically in charge of building up supplies of strategic metals.

After the war, in 1946, Congress passed the Strategic and Critical Materials Stockpiling Act. Because of all the new technology developed during the war, the list of raw materials needed for defense and industry changed and grew, increasing the amount of materials the United States would need to import.

In 1950, Congress allocated $8 million to restock the reserve, continuing the buildup through the decade. By the end of the 1950s, the United States became even more dependent on foreign sources for strategic minerals; all chromium production in the United States ceased and the large Blackbird cobalt mines in Idaho, the major domestic source of cobalt, closed because the low-grade deposits were simply uneconomical to mine any further.

Despite the new threats to supply, the government sold off parts of the stockpile which were deemed unimportant during the 1960s, including aluminum, nickel, copper, and cobalt. Sixty million pounds (60 percent of the amount in the stockpile) of cobalt were sold between 1964 and 1976. Eventually, Presidents Ford and Carter wanted to replenish the stockpile, but Congress would not authorize it. In 1980, President Reagan created a strategic materials task force, the National Materials and Minerals Program Board, to evaluate the U.S. situation. On the recommendations of the board, Reagan ordered stockpile administrators to supplement supplies of 13 materials in 1981. The highest priority was given to cobalt, of which 5.2 million pounds were purchased. Throughout the 1980s, the government continued to strengthen the stockpile.

How the United States Viewed Critical and Strategic Minerals in the 1980s

In the early 1980s, an increased awareness of the importance of strategic minerals and the instability of their supply resulted in public debate over how the precarious the situation really was. Those who believed a crisis was at hand pointed out that supplies were limited and could become inaccessible at any moment, while demand was growing rapidly (in 1985, the Office of Technology Assessment reported that U.S. demand for chromium and platinum group metals was likely to double by the year 2000). They even suggested that foreign sources might form a mineral cartel much like the Arab oil cartel and artificially raise prices or cut off supplies to the United States altogether.

The other side pointed out that the mineral/oil comparison was unfounded: minerals cost much less than oil and require far smaller quantities per unit of output. Besides, most countries supplying minerals to the United States could not afford to halt sales for long. In addition, they claimed that the stockpile was adequate defense against any supply cut-off; even though private industry could not use the government stockpile, most companies which relied on large amounts of strategic minerals had their own 6-12 month back-up supplies. In addition to stockpiles, conservation and recycling could stretch supplies, while new technology could augment them by producing replacement materials and making it possible to extract previously uneconomical ores. Even without a supply crisis, the market possesses an amazing power of self adjustment. United States imports of chromium fell between 1950 and 1970 because of recycling and exporting scrap; at the same time, total world reserves of chromite "rose" 675 percent because of growing technical advances and geological knowledge that were spurred by rising demand and price.

Many people began to look for previously unexplored deposits of critical and strategic minerals, and the sea became one of the most widely known potential sources. A U.S. Geological Survey in 1983 discovered hot springs coming out of ridges on the ocean floor, and the liquids spouting from these springs were rich in minerals. Undersea mountains in the Pacific were found to be covered by crusts rich in cobalt and manganese. The most striking discovery, however, was that metallic nodules found in abundance on the ocean floor contained 30-40 percent manganese, as well as small amounts of cobalt and other metals.

Although the grade of manganese, cobalt, and other ores found in these nodules is better than that found in many land-based mines today, the extreme expenses of ocean mining make their recovery impractical today. However, a rise in price and demand, or a technological innovation that would reduce those costs, could make these nodules and other undersea deposits an economically feasible source of some strategic minerals. Another complication is that in 1982, the United States refused to sign the Law of the Sea, which was designed to govern international seabed mining. The United States wanted countries with heavy mining interests like itself to have greater control over decisions affecting deep-sea mining. The lack of legal certainty over ocean mining also makes the sea an unlikely source of minerals for the time being.

The debate over strategic minerals also brought up the potential reserves of these metals on public lands. For instance, the Absaroka-Beartooth Wilderness in Montana contains reserves of chromium and a platinum-palladium belt along its northern boundary which represents 70 percent of domestic reserves of platinum group metals. Deposits of cobalt near the Blackbird Mine in Idaho extend into the River of No Return Wilderness. Although Congress tried to leave areas rich in minerals out of the River of No Return Wilderness and declared 39,000 acres a Special Mining Management Zone, 1900 claims were staked in the wilderness area before it was closed to exploration. Alaska, home to more wilderness areas than any other state, also contains deposits of cobalt in conjunction with copper and nickel in the south-east, platinum group metals on the western coast, and even chromite in the western Brooks Range.

Those who predicted a strategic minerals crisis claimed that opening up more federally protected wilderness areas to mining was the only option for the United States. They cited the Mining Law of 1872: "all valuable mineral deposits in land belonging to the United States . . . shall be free and open to exploration and purchase under regulations prescribed by law."

However, this law is contradicted and limited by more recent laws. The Endangered Species Act of 1973 prohibits mining operations that would threaten endangered species. Although the Wilderness Act of 1964 prohibited exploration for minerals and patenting of claims on wilderness lands after December 31, 1983, many mining companies established mineral rights on federal lands before that deadline. The act specifies that mining in these wilderness areas must be "substantially unnoticeable," and also prohibits roads, power lines, and mechanical equipment and air transportation to and from the wilderness. Unfortunately, the Bureau of Land Management, which manages many wilderness areas, has tended to interpret the act in favor of mining interests, thus endangering the natural environments the wilderness designation was meant to protect.

Recent presidential administrations have also seemed to favor mining interests. In 1982, the Reagan administration unsuccessfully proposed the Wilderness Protection Act (sometimes called the "Wilderness Destruction Act" by environmentalists), which would have made it easier to re-open federal lands previously withdrawn from mining, especially if the land contains strategic minerals. The act would have set strict deadlines for Congress to designate federal lands as wilderness (and thus protect them from mining), opened wilderness areas to new mining after the year 2000, and reformed environmental legislation that was "burdensome" to the mining industry. In addition, the Reagan administration tried to extend the period for staking mineral claims 20 years past 1983. Although the act was not passed, such attempts reveal that public lands will always be vulnerable to exploitation by mining interests as long as an economic demand for the minerals on those lands exists.

Despite the long-standing nature of the conflict between environmentalists and those who favor mining, it is not irreconcilable. In the Rocky Mountains, where many wilderness areas are rich in minerals, only 4 percent of the land is actually designated as wilderness. Many other mineral deposits, on private and nonwilderness federal land, are available to be developed without jeopardizing the wilderness areas. In addition, some mining operations have avoided wilderness destruction by locating mine entrances outside wilderness areas, even when the mine extends underneath the wilderness.

LOOKING AHEAD: THE FUTURE OF CRITICAL AND STRATEGIC MINERALS IN THE UNITED STATES

Minerals are a nonrenewable resource, and thus both domestic and foreign supplies are finite. Although advances in mining technology may make it possible to extract previously uneconomical ores in the future, global demand for most major minerals is predicted to double during the 1990s because of the growing human population and the rising standard of living in many nations. Thus, our supply of many minerals may well be exhausted someday, even if we expand mining into controversial sites such as domestic wilderness areas. Debates over mining development versus wilderness preservation are really only tangential to the effort to secure a long-term, stable supply of critical minerals. In the end, we will have to solve the real strategic minerals problem through recycling, conservation, and developing new, substitute materials.

In 1982, Congress allocated money for conservation and substitution efforts in addition to strategic material production. Although we currently derive 10 to 15 percent of our chromium supply from recycling, the Office of Technology Assessment estimates that we could add another 3.5 percent annually with stronger efforts. Significant amounts of cobalt are currently lost or downgraded as scrap and waste, and recent research has investigated new ways to extract cobalt from lead smelter waste. Although manganese is difficult to recycle, improvements in steel processing could reduce the U.S. demand by 45 percent by the year

2000. The Office of Technology Assessment predicts that consistent recycling of catalytic converters could recover 400,000 - 500,000 troy ounces of platinum group metals (approximately 20 percent of 1980 consumption level) annually by the mid-1990s.

Advanced production technology could also "expand" supplies of critical and strategic minerals. For example, Pratt & Whitney has developed a method that uses less metal in jet engines, minimizing the amounts of strategic metals needed. Previously, huge chunks of metal were slammed by molds, wasting large amounts of metal. The new process, called gatorizing, heats small bars of metal until they are soft and then presses them between the two molds. Gatorizing eliminates the amount of wasted metal by more than half. In addition, General Electric has found a way to use less heat-resistant cobalt in engine blades by drilling tiny holes in the blades to encourage air-cooling.

Another promising area is secondary substitution. For instance, instead of trying to make stainless steel without chromium, manufacturers would try to produce tools or furniture without stainless steel.

Primary substitution has made some exciting advances, as well. One field that promises to help produce substitutes for strategic metals is advanced ceramics. Recently developed ceramic materials, which can withstand higher temperatures and more hostile environments, are gaining use in cutting tools, seals, bearings, and sandblasting nozzles. Cutting tools alone consume 20 percent of the cobalt used in the United States. In addition, Pratt and Whitney is experimenting with jet engine combusters (the fuel-burning chamber) that are made out of nickel with a heat-resistant ceramic coating rather than cobalt. Other applications for ceramics in automobile engines which are currently being explored could reduce our demand for chromium by 10 percent and our consumption of platinum by as much as 50 percent.

Plastics are another category of potential substitutes. Newly developed plastics can withstand temperatures of 500° C (932° F), and have been used in rotor blades of gas turbine engines. Researchers are also currently working on plastic auto engines. Although auto engines contain only small amounts of critical materials, any reduction in their use would be significant when multiplied by the 7 million automobile engines produced in our country every year.

Carbon-carbon composites are a new material being developed in hopes of replacing superalloys in jet engines. They consist of carbon fibers held together by a resinous matrix; when the matrix dries, the material becomes a solid compound.

Although researchers and industry are making advances in conservation and substitution, the current low prices and steady supply of critical and strategic minerals do little to encourage serious efforts toward eliminating U.S. dependence on these minerals, whether from foreign or domestic sources. The U.S. government needs to take responsibility for planning this country's long-term response to the strategic mineral situation by providing funding and other incentives for developing more efficient processes and reliable substitutes.

SUGGESTED READING

Anderson, Ewan. *Strategic Minerals: The Geopolitical Problems for the United States*. New York: Praeger, 1988.

Barbara, Robert J. *Cobalt: Policy Options for a Strategic Mineral*. Washington, DC: Congress of the United States, Congressional Budget Office, 1982.

Betteridge, W. *Cobalt and its Alloys*. New York: Halsted Press, 1982.

Chromium. Geneva: World Health Organization, 1988.

Clark, Joel P., and Frank R. Field III. "How Critical Are Critical Materials?" *Technology Review*. August/September 1985. 39-46.

Crockett, R.N., Gregory R. Chapman, and Michael D. Forrest. *International Strategic Minerals Inventory Summary Report, Cobalt*. Reston, VA: Department of the Interior, U.S. Geological Survey, 1987.

DeHuff, Gilbert L. *Manganese*. Washington, DC: Bureau of Mines, U.S. Department of the Interior, 1979.

Fierman, Jaclyn. "Cutting Dependence on Strategic Metals." *Fortune*. July 22 1985. 69.

Foster, Russell J. *Technological Alternatives for the Conservation of Strategic and Critical Minerals— Cobalt, Chromium, Manganese, and Platinum-Group Metals: A Review*. Washington, DC: Department of the Interior, Bureau of Mines, 1985.

Jacobsen, D.M., R.K. Turner, and A.A.L. Challis. "A Reassessment of the Strategic Materials Question." *Resources Policy*. June 1988. 74-84.

Jones, Thomas S. *Manganese*. Washington, DC: Bureau of Mines, U.S. Department of the Interior, 1983.

Kirk, William S. *Cobalt*. Washington, DC: Bureau of Mines, U.S. Department of the Interior, 1985.

Manheim, F.T. "Marine Cobalt Resources." *Science*. May 2 1986. 600-608.

Matthews, Olen Paul, Amy Haak, and Kathryn Toffenetti. "Mining and Wilderness: Incompatible Uses or Justifiable Compromise?" *Environment*. April 1985. 12-35.

Nriagu, Jerome O., and Evert Nieboer, eds. *Chromium in the Natural and Human Environments*. New York: Wiley, 1988.

Papp, John F. *Chromium*. Washington, DC: U.S. Department of the Interior, Bureau of Mines, 1985.

Plotkin, Rhoda. "The United States and South Africa: The Strategic Connection." *Current History*. May 1986. 201-205.

Ridgeway, James. "Stalking Strategic Metals." *Science Digest*. February 1983. 42-43.

Roberts, Leslie. "Uncertain Prospects for Deep Ocean Mining." *BioScience*. January 1983. 14+.

Robinson, Arthur L. "Congress Critical of Foot-Dragging on Critical Materials." *Science*. October 3 1986. 20.

Shafer, Michael. "Mineral Myths." *Foreign Policy*. Summer 1982. 154-171.

Strategic Materials: Technologies to Reduce U.S. Import Vulnerability. Washington, DC: U.S. Congress, Office of Technology Assessment, OTA-ITE-248, May 1985.

Weston, Rae. *Strategic Minerals: A World Survey*. London: Rowman and Allanhead, 1984.

Willenson, Kim. "The Mines of Apartheid." *Newsweek*. August 11 1986. 30.

Young, Gordon. "The Miracle Metal: Platinum" *National Geographic*. November 1983. 686-706.

CHAPTER *15*

NUCLEAR RESOURCES: THE FEDERAL NUCLEAR RESERVATION AT HANFORD, WASHINGTON

Less than fifty years ago, the idea of tapping into the energy that binds together the nucleus of an atom was purely theoretical, the stuff of complex physics theories and science fiction novels. Then, during World War II, fiction became reality. In an effort to develop nuclear weapons before Nazi Germany, the United States initiated the Manhattan Project, a top-secret effort to create the world's first nuclear bomb. In 1945, Manhattan Project scientists succeeded in developing the bomb before the Germans, but ironically, the United States never used its new weapon against Hitler's regime. Instead, in the early morning hours of August 6, 1945, a U.S. warplane named the Enola Gay released the first atomic bomb on the city of Hiroshima, Japan. The ensuing destruction clearly demonstrated that humankind was capable of destroying life on a scale previously unimagined.

Although the atom's awesome potential for generating energy was first used in such a destructive manner, nuclear researchers soon began to develop other applications for nuclear resources, including the generation of electricity and the treatment of cancer patients. To most people at that time, the potential of nuclear power seemed limited only by our ability to conceptualize its possible uses and benefits. However, as the nuclear industry developed, scientists and laypeople alike became more and more aware of the unanticipated and possibly uncontrollable dangers of nuclear power.

Today, our society's use of nuclear resources remains an issue of great controversy. Some say nuclear resources are indispensable to our society: we need nuclear weapons to prevent other nations from using nuclear weapons against us, and nuclear power provides a desirable alternative energy source to fossil fuels because it does not contribute to the greenhouse effect or acid precipitation. However, others point out that nuclear resources possess the potential to end life on earth as we know it. Even if we can avoid all-out nuclear warfare or radiation-releasing accidents at nuclear power plants, we have no safe, permanent way to deal with the vast amounts of radioactive waste that the nuclear energy and weapons production industries produce—waste that can remain deadly for millions of years.

Do the tremendous benefits of nuclear resources outweigh its risks? We as a society must decide the answer to this question. However, no matter what course of action we choose, we must not repeat our past mistakes in dealing with nuclear resources. Our country's history of managing nuclear energy has been characterized by haste and irresponsibility, and the nuclear weapons industry in particular has proceeded with a flagrant disregard for human health and the environment. Like the commercial nuclear power industry, the nuclear weapons industry is characterized by the risk of a reactor accident and the dilemma of waste storage. In addition, those responsible for producing nuclear weapons have historically used the name of national security to keep secret both deliberate and accidental contamination of air, soil, and water by potentially deadly radiation.

In this chapter, we will explore the explosive issues involved in the development of nuclear resources by focusing on one site in the nuclear weapons production chain. During World War II, the United States took over 570 square miles in southeastern Washington in order to establish the Hanford Reservation, a facility for producing plutonium to fuel the nuclear weapons developed as part of the Manhattan project. Hanford plutonium powered the world's first atomic explosion—the "Trinity" test in Alamogordo, New Mexico—and the bomb dropped on Nagasaki, Japan.

Activities at Hanford eventually expanded beyond plutonium production into areas such as electrical power generation and research and development. However, today almost all these activities have ceased because of a decreased demand for plutonium, a growing concern about safety, and decades of inadequate waste disposal practices. The general public has only recently learned the truth about the numerous ways operations at Hanford have endangered the surrounding environment and human population. In addition, today many experts believe that Hanford is the most contaminated site in our country's nuclear weapons production complex. Although the future of the facility is uncertain, it must include finding a way to minimize the risks that Hanford's activities and wastes pose to the surrounding area.

PHYSICAL BOUNDARIES: GEOLOGY AND GROUNDWATER FLOW OF THE HANFORD SITE

The Hanford site lies on a semi-arid desert plain, mostly flat except for two small east-west ridges, Gable Butte and Gable Mountain. The Rattlesnake Hills bound the area to the south, the Yakima and Umtanum Ridges rise along the western edge, and the Saddle Mountains lie to the immediate north. The Columbia River flows through the northern portion of the reservation and then forms most of its eastern border. Downriver from Hanford lie the "Tri-Cities" of Richland, Kennewick, and Pasco, with a combined population of about 144,000.

The geology of this area has some unique characteristics that affect the movement of groundwater, some of which has been contaminated by radioactive wastes produced by operations at Hanford. The ground beneath Hanford is made up of three major layers. Directly below the soil lies a gravel layer left behind by receding glaciers. Water and other liquids percolate readily through these gravels until they reach a sedimentary layer, deposited by the Columbia River before the last Ice Age. The water table in the Hanford area generally coincides with the top of this sedimentary layer. Because the third layer of basalt, or hardened lava, is relatively impermeable, groundwater usually moves slowly eastward through the dense sedimentary rock, eventually entering the Columbia through silver-dollar sized springs.

However, in some areas water can move much more quickly. For millions of years before the glaciers deposited the gravel layer, wind and retreating floodwaters carved channels in the exposed sedimentary rock—channels that were later filled by the gravel. Some of these gravel-filled channels lie below the water table, funnelling groundwater to the Columbia much more quickly than it could travel through the denser sedimentary rock.

BIOLOGICAL BOUNDARIES: THREATS TO THE ENVIRONMENT AND TO HUMAN HEALTH

Besides threatening to spread radioactive waste and other hazardous material through the groundwater, the activities at Hanford Reservation pose other potential threats to the area's biotic community.

The most predominant form of vegetation in the Hanford area is a desert shrub known as sagebrush, or tumbleweed. Cheatgrass and bluegrass are also common. Because such plants can absorb radioactive substances from the ground, water, and air, they can pass them along the food chain. In addition, sagebrush has caused additional problems by spreading radioactive contamination off-site and across the reservation as it tumbles in the wind. The long tap root of the sagebrush (15 feet) provides a ready pathway for the uptake of radioactive liquids in an arid environment. To alleviate this problem, Hanford regularly sprays defoliants to prevent plant growth on contaminated soil sites.

The Hanford area's vegetation supports a broad range of insects, birds, and larger animals, including elk, mule deer, coyote, rabbits, mice, ground squirrels, pocket gophers, raccoons, and porcupines. Near the Columbia River, bald eagles make their nests in the sparse tree population. The river area is also a major resting area for migratory waterfowl. Furthermore, the river is home to a wide variety of aquatic life, ranging from plankton to migratory fish such as the Steelhead trout and the Chinook, Sockeye and Coho salmon. Although these fish are not as likely to accumulate significant concentrations of radioactive contamination as resident species such as carp, bass, perch, and catfish, which spend their whole lives in the Hanford stretch of the Columbia, their tremendous recreational and commercial value makes their fate a matter of great public concern. Although Native Americans in the area feel that Hanford's operations have played a large role in the decline of migratory species, Hanford officials claim that other factors, such as dams, siltation, and sewage, have been responsible for the species' decreased numbers.

Radioactive substances accumulated in plants and animals in the Hanford area could potentially move on to the human community through the food chain; fish from the Columbia and milk from cattle that graze in the surrounding area are common paths of contamination. In addition, human exposure can occur through drinking water and airborne radiation. Radiation can have a wide range of effects on human health, depending on the level of exposure. Extremely high levels of exposure—such as those produced by a nuclear explosion— can kill immediately or cause radiation sickness. Slightly lower levels can cause radiation burns, destroy bone marrow, and damage internal organs.

Even the lower levels of exposure more typical of the Hanford area pose health hazards because these small amounts of radiation can cause cell damage that may accumulate over time. These damaged cells continue to grow, creating the possibility of cancer years later. Some scientists believe that any amount of radiation, no matter how minute, affects biological systems in some way, even though it could be so slight as to be undetectable.

Unborn children are especially likely to suffer severe effects from radiation exposure. If doses are high enough, a fertilized egg can develop lesions that will prevent its implantation in the uterus, thus causing a spontaneous abortion. Lower doses of radiation allow the fertilized egg to implant, but can damage the fetus' developing organ systems. Such fertilized eggs can develop into live-born children with mild to severe birth defects.

SOCIAL BOUNDARIES: THE EFFECT OF THE CULTURAL CLIMATE ON THE HANFORD SITE

The Hanford area was originally inhabited by Native Americans, who ceded most of their territory to the federal government in 1855. Although this native population retained the right to fish, hunt, gather roots and berries, and graze horses and cattle on the land, white farmers and ranchers gradually moved into the area and used it for orchards, crops, and grazing.

However, World War II radically altered the character of the area in 1943, when the Manhattan Project designated Hanford as the site for a top secret effort to produce plutonium, which would fuel nuclear weapons. Hanford seemed an ideal site for this effort because of the sparse population, the ample supply of water, provided by the Columbia River, for cooling the reactors, and the easy access to railroad systems. Approximately 1500 people who had been living on the proposed site were relocated. The large number of employees needed by the facility encouraged the growth downstream of the Tri-Cities area, which eventually became the fourth largest metropolitan area in the state.

Unfortunately, the lax attitude toward environmental safety common at the time allowed construction of the facility to continue for over a month before investigations into groundwater and subsurface conditions were initiated. Disregard for safety was compounded by the understanding that operations at Hanford were only temporary; they would cease after the war. After World War II ended, however, plutonium production actually increased because of the arms build-up of the Cold War. By 1964, six other reactors had joined the original three built in 1944.

LOOKING BACK: THE HISTORY OF THE HANFORD NUCLEAR RESERVATION

Hanford was once a major link in the military's chain of plutonium production facilities across the country. Hanford's mission was to transform uranium 238, which had been processed and enriched at other government sites, into plutonium 239, the raw material for nuclear bombs. Inside the reactors at Hanford, rods of uranium fuel were bombarded with neutrons, which caused atoms of the uranium to split, or fission. In addition to producing radioactive energy, the process of fission released new neutrons from the split nuclei. Many of these neutrons bombarded other uranium atoms, resulting in a sustained chain reaction. Other neutrons were absorbed by atoms of uranium 238, transforming them into plutonium 239. The highly radioactive substances produced by the process of fission are collectively known as fission products.

Once the uranium fuel of a Hanford reactor was sufficiently irradiated, it was then removed to the Plutonium-Uranium Extraction (PUREX) facility, where it was reprocessed.

Reprocessing involved dissolving the spent fuel in a highly acidic solution in order to separate plutonium and uranium from other fission products. The plutonium thus extracted was then shipped from Hanford to other facilities, where it was fashioned into hydrogen bomb triggers.

Today, plutonium production at Hanford has come to a virtual standstill. Eight of the nine plutonium production reactors were deactivated between 1964 and 1971 because of decreasing demand for plutonium. The ninth— the N reactor—operated alone after 1971 until it was placed on standby in 1988 because of safety concerns. The PUREX facility stayed in operation after the N reactor ceased production. In fact, PUREX had a big enough backlog to continue extracting plutonium and uranium until around 1995. However, after PUREX experienced steam pressure problems in late 1988, it was shut down, and public concern over the highly radioactive waste it produces has kept it inoperative pending further study.

With improved relations between the United States and former communist-block countries, it seems more and more unlikely that plutonium production for nuclear weapons will once again become a major national defense priority. Even if it does, the age and outdated design of much of the Hanford equipment would probably rule out using it as a major plutonium production center in the future. However, we still face the question of what to do with Hanford and other nuclear facilities across the country which are plagued with aging, potentially unsafe equipment, huge quantities of highly radioactive waste, and contaminated soil and groundwater.

In addition, we need to examine national policies that have allowed government nuclear weapons facilities to function with little or no regard for the surrounding environment or the health of workers and nearby residents. Until very recently, the Department of Energy (DOE) and its predecessor, the Atomic Energy Commission (AEC), managed military nuclear facilities in almost complete secrecy. The Atomic Energy Act, passed in 1954, granted the AEC the right to manage weapons productions, with no regulation or oversight by any other government agencies, for reasons of national security. Unfortunately, the AEC, and later the DOE, took advantage of this lack of oversight to conceal potentially dangerous leaks and other releases of radioactive substances from the public. In addition, the special status of the DOE has rendered Hanford and other federal weapons facilities exempt from civilian regulations on radioactive waste management, environmental protection, and worker safety.

Clearly, we must eliminate the threat that Hanford poses to the surrounding environment. To accomplish this goal, however, we must ask how such a massive undertaking should be managed and what will happen to the people of the surrounding communities, who rely on the Hanford facility as a source of employment? In addition, we need to closely examine the national policies that allowed such dangerous situations to develop in the first place. However, in order to determine how best to manage Hanford and the larger issue of nuclear weapons production in the future, we first need to fully understand the history of the Hanford site.

How Plutonium Production Has Affected the Hanford Environment

During Hanford's early years, radioactive wastes were disposed of with little accurate knowledge as to how they would affect the environment. Even as knowledge improved, environmental threats persisted because of the sheer volume of waste produced; over 200 billion gallons of waste were discharged to the environment from Hanford reprocessing plants between 1945 and 1985. Experts agree that Hanford is the most heavily contaminated of all U.S. government nuclear facilities.

Even more shocking is the fact that virtually all of the risks Hanford operations posed to human health and the environment were kept secret from the general public. Not only had plant officials never warned residents about any of these risks, but they had also repeatedly assured residents that the plant's operations were harmless. Despite the DOE's policy of secrecy, the truth was finally revealed in February 1986, as the result of an extensive Freedom of Information Act request filed by two public interest watchdog groups: the Environmental Policy Institute and the Hanford Education Action League. As the result of this and other requests, the DOE released a total of 45,000 pages of previously withheld documentation regarding past operations at Hanford. This documentation revealed that the facility had released billions of cubic meters of radioactive gases into the air and discharged billions of gallons of radioactive liquids into the Columbia River and the surrounding ground.

Airborne Emissions. Although early operations at Hanford were extremely lax in terms of environmental contamination of all kinds, the earliest serious contamination came from radioactive emissions to the air. Emissions from the stacks of reprocessing plants carried approximately 536,000 curies of radioactive iodine into the environment over the first 12 years of Hanford operations (1944-1956). Between December 1944 and November 1946 alone, the plants released an estimated 470,000 curies. In 1949, plant officials purposely emitted high levels of radioactive iodine as part of an experiment designed to lead to better monitoring of Soviet tests of atomic weapons. Studies of the area's vegetation performed immediately after that release revealed contamination levels up to 600 times greater than the accepted limit.

These emissions are believed to be responsible for the high rate of cancer of the thyroid gland among Hanford area residents. Radioactive iodine concentrates in the human thyroid gland, where it can cause cancer and general hormonal dysfunction. Residents are believed to have taken in the iodine not only from the air, but also by drinking milk from cows and goats that had grazed on contaminated vegetation. A Washington State Department of Social and Health Services estimated that in 1945, infants in nearby Pasco consuming local milk might have received radiation doses to the thyroid gland at levels as high as 2300 rem per year. Infants as far away as Spokane would have received doses of up to 256 rem that year. By comparison, current health studies suggest that radiation doses of 20 rem or less can cause damage to the thyroid gland, and Washington state law today restricts doses of radiation to the thyroid gland and other critical organs to .075 rem per year.

Contamination of the Columbia River. The nearby Columbia River has also been a frequent victim of inadequate waste disposal at Hanford, passing on contamination to the plants, animals, and humans that rely on its waters. Between 1943 and 1964, cooling water from the first eight plutonium production reactors was dumped directly into the river after an initial period of decay. Although cooling water is considered low-level waste, remaining dangerous for fairly short periods of time, government studies performed in 1947 revealed that its disposal into the Columbia was potentially dangerous. These studies found that tissues of fish, crustaceans, plankton, algae, and other aquatic life contained concentrations of radioactivity, averaging 10,000 times that of the river water, apparently because of accumulation through the food chain.

Contamination levels increased throughout the 1950s, as five new reactors were added to the three built in 1947. In addition, dangerous concentrations of radioactive substances meant to be contained on the reservation were allowed to reach the river because of careless disposal practices and an incomplete knowledge of Hanford geology.

Up until 1973, liquid wastes—including toxic chemicals, low-level radioactive waste, and transuranic waste containing highly radioactive, human-made elements such as plutonium—were drained directly to the soil at Hanford. Originally, officials did not worry about such wastes reaching the Columbia because they believed it would take the liquid at least fifty years to travel to the river. Over that time, they believed that the most dangerous radioactive compounds would decay or be filtered out of the liquid by the soil. However, they were unaware of the gravel-filled channels in the sedimentary rock below the water table. In 1963, Hanford scientists found that radioactive tritium from cooling water used at the PUREX plant that had opened in 1956 had already traveled to the Columbia—just seven years after it was released. Groundwater monitoring reports showed that in some areas of the river, concentrations of tritium were above EPA drinking water standards.

Although direct dumping ceased around 1973, most low-level liquid wastes were still deposited in ponds, trenches, ditches, and cribs, from which they could quickly drain into the soil. Once in the soil, the radioactive substances did not always decay or become absorbed by the soil. In some cases, the soil became so saturated with the compounds that it was unable to absorb any more. In 1977, Hanford scientists realized that unusually high concentrations of radioactive strontium 90 were entering the Columbia through a small spring about 800 feet from a crib containing low-level liquid waste from the N reactor. In that year alone, releases of strontium 90 to the river tripled, although there was no corresponding increase in the amount of liquid being drained to the crib. The soil's ability to absorb strontium 90 had clearly been exhausted.

Contamination of Soil and Groundwater. In addition to contaminating the Columbia, liquid wastes from Hanford pose major threats to the area's groundwater. Perhaps the most highly publicized of these threats has been the inadequate disposal of high-level waste, which remains radioactive for tens of thousands of years or more. A great deal of Hanford's high-level waste has been produced by PUREX and earlier reprocessing facilities, which separate usable uranium and plutonium from fission products such as radioactive isotopes of iodine and a radioactive isotope of hydrogen called tritium. During its years of operation, PUREX produced one million gallons of this waste annually.

The AEC originally believed that such high-level wastes could be effectively stored in single-shell carbon steel tanks buried a few meters below ground. Between 1943 and 1980, Hanford officials stored 46 million gallons of high-level waste in 149 single-shell carbon steel tanks. Although these tanks were only intended for temporary storage, it was estimated that they could last up to 300 years if necessary. But in 1956, after only 13 years, Hanford officials found that a total of 29 tanks were leaking, and another 31 tanks were categorized as "suspected leakers."

Many of the leaks stemmed from efforts to prevent the corrosive wastes from dissolving the tanks. Lye, added as a neutralizing agent, also caused the radioactive elements to precipitate as sludge at the bottom of the tanks. The intense concentration of heat produced by the sludge cracked some of the tanks, spilling an estimated 500,000 gallons of high-level waste into the soil.

Although undamaged single-shell tanks still contain 46 million gallons of high level waste, 28 sturdier double-shelled tanks constructed in 1968 contain about 14 million gallons as well. Unfortunately, the new tanks are not an ideal solution either; pits of rust have been found in their stainless steel liners.

In the past, groundwater has been contaminated because of a combination of inaccurate records of past waste disposal and lack of attention to the way certain wastes would react with substances in the soil. In July 1984, liquid waste from uranium extraction operations was routed into a newly constructed crib. Similar in design to a septic tank drain field, a crib depends on soil to trap and filter radioactive materials. Within a few months, uranium concentrations in groundwater monitoring wells near the new crib jumped dramatically to 170 times their previous levels and more than 5000 times the current EPA drinking water standards. Experts feared that once in the groundwater, this contamination could easily spread to the Columbia and surrounding communities.

Scientists found that the liquid deposited in the new crib had percolated downward over 100 feet until it ran into a thick crust of calcium carbonate, which forced it to move horizontally to an area below two older, abandoned cribs approximately 265 feet away. These cribs had received large quantities of uranium, along with a small quantity of acidic decontamination waste, before closing in the late 1960s. The acidic decontamination waste had partially dissolved the uranium compounds, enabling them to slowly drift downward, until they interacted with the calcium carbonate layer, forming even less stable uranium carbonate complexes. The liquid from the new crib picked up the uranium in the calcium carbonate layer and carried it down into the water table through a forgotten and unmarked reverse well through which liquid wastes had been pumped deep into the ground.

Although Hanford's waste contractor prided itself on the million-dollar effort to clean up after the "unique" incident, uranium levels remained more than 1300 times greater than the EPA's drinking water standard.

Even though all the liquid waste produced at Hanford did not carry contamination into the groundwater, the approximately 30 billion gallons of low-level liquid waste dumped into the ground have left Hanford's soil widely and heavily contaminated. Some experts estimate the amount of plutonium 239 discharged to the soil alone would be enough to construct 40 bombs of the type used on Nagasaki. When the facility is cleaned up, all of the contaminated soil will have to be removed and stored, vastly multiplying the amount of radioactive waste to be dealt with.

Ongoing Risks. Some area residents and environmental groups are worried that the DOE is compounding contamination problems at Hanford by using it to store outside waste as well as the waste produced on the reservation itself. They fear that the area will become a "nuclear junkyard" contaminated beyond repair. Four 1000-ton reactor sections from six retired Polaris nuclear-powered submarines are already buried on the premises. The navy plans to retire approximately 100 more of these subs over the next 20 to 30 years, and Hanford, which has the space to accommodate them, could easily become a targeted disposal site. Also interred on Hanford's grounds is the country's first commercial nuclear reactor, the Shippingport Atomic Power Station.

Although public dissatisfaction with Hanford's history of waste contamination and accidental leaks is steadily growing, a new area of concern surfaced when the world learned of the disaster at Chernobyl. In many respects, the 23-year-old N reactor at Hanford bore an uncanny resemblance to the ill-fated Soviet plant. The designs of both reactors allowed them to simultaneously produce plutonium for military use and steam for generating electricity. More importantly, both used graphite to moderate the rate of fission—the N reactor is the only U.S. reactor to use the dangerously flammable substance that caught fire and spread radioactive contamination during the Chernobyl accident. In addition, the N reactor used a more volatile form of uranium fuel than the Soviet reactor.

Both reactors also lacked the customary high-density concrete containment structures that guard against the release of radioactivity to the environment in case of an accident; again, the N reactor is the only major reactor facility in the United States without such a containment structure. In fact, the N reactor's containment limit was only 5 pounds per square inch, substantially less than Chernobyl's 27 pounds per square inch limit.

Despite efforts by DOE officials to convince the public that a major accident was unlikely to occur at Hanford, the public as well as environmentalists called for an immediate investigation of N reactor safety. In late 1986, six consultants to the DOE recommended a six-month, $50 million safety overhaul for the plant, and the plant was closed down pending repairs. However, lack of attention to important safety features such as a containment shield led environmental organizations to announce that they would file suit for a full environmental impact statement if DOE tried to restart the facility after the repairs. The N reactor was never reactivated, and it is currently in the process of being decommissioned, or retired.

How the Government Plans to Clean Up Hanford

In response to legislation passed by Congress, in 1983 the Department of Energy proposed the Defense Waste Management Plan, a blueprint for the clean-up and long-term waste management of its major nuclear sites. The plan included burying transuranic wastes at a nuclear waste repository planned in New Mexico; solidifying radioactive sludge from high-level tanks into glass (vitrification) and burying it in a deep repository planned for Nevada; and grouting low-level waste, or mixing it with cement and pouring it into concrete-lined pits.

However, the DOE was slow to implement this plan. When Congress reauthorized the Resource Conservation and Recovery Act (RCRA) in 1984, states received new authority over solid and hazardous liquid waste problems. Responding to the lack of cleanup effort by DOE, the Washington Department of Ecology decided to confront the Hanford facility on its failure to comply with state standards. However, when Ecology identified 20 violations, DOE pointed out that almost all the wastes at Hanford were "mixed" —they contained radioactive as well as hazardous substances. DOE claimed that RCRA did not apply to mixed wastes because the Atomic Energy

Act gave DOE complete control over any waste containing radioactive substances produced by federal facilities. Although the EPA insisted that RCRA covered mixed wastes as well as simply hazardous ones, the law itself was inconclusive.

Several members of Congress proposed bills that would have given the EPA definite jurisdiction over mixed wastes, but they did not pass. However, in 1986, the State of Tennessee won a lawsuit against a DOE plant in Oak Ridge, setting a precedent for applying

RCRA to mixed wastes. Soon after, Washington's Department of Ecology fined DOE for its violations and, along with EPA, demanded compliance on the five worst violations. DOE agreed to comply but not to pay.

On May 15, 1989, Ecology, EPA, and DOE came together in the Tri-Party agreement, an unprecedented clean-up program that calls for spending $50 billion over next 30 years to eliminate Hanford's worst problems. The EPA is responsible for enforcing clean-up of old waste, while the state is involved in controlling the wastes being currently produced. At Hanford, both contaminate the same soil and groundwater, and the Tri-Party agreement was intended to work out which agency would "call the shots" at which waste site, thus eliminating duplication of efforts and conflicting decisions. The agencies estimate that they will need $2.8 billion in the next five years to hasten removal of liquids from single-shell tanks; study how to remove radioactive sludge from the bottom of the tanks; install new groundwater monitoring wells; investigate old waste sites; and begin to grout and vitrify waste from double-shell tanks.

However, the removal of liquids from the single-shelled tanks will not happen within five years, if ever (some of the chemicals in the tanks become explosive when dry or hot). Also, DOE is dragging its feet in researching methods of removing the solids from the tanks. Grouting has been delayed two years and DOE is expected to announce shortly that there will be further delays. In December 1991, DOE and Washington's Department of Ecology agreed to build a waste vitrification plant at Hanford. The plant should be operational by December 1999. However, the DOE does not plan to use it to vitrify most of the waste from the single-shelled tanks, prompting critics to claim that the DOE is not serious about cleaning up waste.

Although the Tri-Party agreement is overall a positive step, critics note that the agencies involved have no way of ensuring that Congress will continue to appropriate funds for the clean-up over the years. In addition, many people at public hearings strongly objected to a provision that would allow PUREX to continue operating until 1995. As a result, the three agencies agreed to a separate, 14-month investigation of liquid waste from PUREX and other facilities to determine if some or all of the discharges should be stopped or phased out.

LOOKING AHEAD: THE FUTURE OF
THE HANFORD NUCLEAR RESERVATION

With the N reactor in the process of being decommissioned and the PUREX facility closed and under investigation, it seems unlikely that Hanford will continue plutonium processing operations. Even the DOE, in its projections for nuclear facilities, foresees that Hanford will be out of production in the year 2010. But what will happen to the facility?

In addition to the massive clean-up initiated by the Tri-Party agreement, all nine nuclear reactors and PUREX will eventually need to be decommissioned. These facilities contain vast amounts of radiation that must be prevented from contact with humans or the environment for up to 4.5 billion years. Disposing of them safely could cost hundreds of millions of dollars.

A different fate has been proposed for the B reactor; some suggest making this facility, which produced the plutonium for the bomb dropped on Nagasaki, into a museum because of its key role in ending World War II. Others question the safety of making a nuclear facility a public monument as well as the appropriateness of glorifying the death and destruction caused by the plutonium-powered bombs.

Although Hanford will eventually become a costly, contaminated reminder of our carelessness and ignorance in handling nuclear energy, the mistakes that occurred there can still serve a positive function: preventing a similar scenario in the future. Grass roots groups like HEAL insist that the only way to prevent continued abuses of human health and the environment by the nuclear weapons industry is to limit the DOE's right to secrecy in the name of national defense. The public should have access to detailed reports of the way defense agencies and even military contractors handle radioactive and other hazardous substances. Although DOE claims to have adopted a new attitude about environmental and health concerns, some workers at the Hanford site state that, upon reporting safety problems to their superiors, they have been harassed and even fired.

Individual states such as Washington are also insisting that the DOE reform its practices. These states want Congress to pass legislation that will leave no doubt that all federal agencies must comply with same laws as other businesses and institutions, especially when it comes to human and environmental safety. Laws that would work toward this goal include one the House passed in July 1989, allowing the EPA and state hazardous waste enforcement agencies to penalize the DOE for violating RCRA. This law would also waive the DOE's immunity from complying with other state and federal environmental regulations. Another proposed law, the Federal Nuclear Facilities Environmental Response Act, would establish a trust to finance clean-up efforts, decommissioning processes, environmental compliance renovations, and long-term monitoring of federal facilities. The act would also create a joint EPA-DOE research and development program focusing on new nuclear compliance and clean-up technologies.

The "nuclear age" that we have lived in since World War II has left a legacy of dangerously contaminated areas like Hanford. Our still inadequate knowledge of nuclear energy and its effects is one barrier to cleaning up these facilities, but a much stronger one is our government's lack of accountability for its nuclear energy policies and their results. Only when we force them to take responsibility will we begin to work towards an environment safe from radioactive waste and accidents.

SUGGESTED READING

Charles, Daniel. "The People vs. the Complex." *Bulletin of the Atomic Scientists.* January/February 1988. 29-30.

Connor, Tim. *Hot Water: Groundwater Contamination at the Hanford Nuclear Reservation.* Spokane, WA: Hanford Education Action League, 1986.

Fradkin, Philip L. *Fallout: An American Nuclear Tragedy.* Tucson: The University of Arizona Press, 1989.

Franklin, Karen E. "It's Only Plutonium." *The Progressive.* October 1991. 15-20.

General Accounting Office. *Nuclear Waste: Unresolved Issues Concerning Hanford's Waste Management Practices.* Washington, DC: 1986.

Gray, Robert H. "The Protected Area of Hanford as a Refugium for Native Plants and Animals." Environmental Conservation. Autumn 1989. 251-260.

Gray, R.H., R.E. Jaquish, P.J. Mitchell, and W.H. Rickard. "Environmental Monitoring and Hanford, Washington, USA: A Brief Site History and Summary of Recent Results." Environmental Management. September/October 1989. 563-572.

Levi, Barbara G. "Hanford Seeks Short- and Long-Term Solutions to Its Legacy of Waste." Physics Today. March 1992. 17-21.

Kunreuther, Howard, William H. Desvousges, and Paul Slovic. "Nevada's Predicament." Environment. October 1988. 16-33.

Marshall, Eliot. "Hanford's Radioactive Tumbleweed." Science. June 26 1987. 1616-1620.

National Academy of Sciences. Radioactive Wastes at the Hanford Reservation: A Technical Review. Washington, DC: 1978.

Patterson, Walter C. The Plutonium Business and the Spread of the Bomb. London: Wildwood House (for the Nuclear Control Institute). 1984.

Peterson, Cass. "The Bomb-Business Blues." Sierra. May/June 1988. 33-38.

Raloff, Janet. "Hanford Reactor's Safety is Questioned." Science News. August 16 1986. 101-102.

Renner, Michael G. "War on Nature." World Watch. May/June 1991. 18-25.

Russell, Dick. "In the Shadow of the Bomb: Cleaning Up after DOE." The Amicus Journal. Fall 1990. 18-30.

Saleska, Scott, and Arjun Makhijani. "Hanford Cleanup: Explosive Solution." The Bulletin of the Atomic Scientists. October 1990. 14+.

Shulman, Seth. "Hanford Nuclear Radiation Doses Assessed." Nature. July 19 1990. 205.

——. "Nuclear Power: The Dilemma of Decommissioning." Smithsonian. October 1989. 56-66.

——. "Nuclear Waste: New Fears at Hanford." Nature. August 9 1990. 501.

——. "When a Nuclear Reactor Dies, $98 Million is a Cheap Funeral." Smithsonian. October 1989. 56+.

Steele, Karen Dorn. "Hanford: America's Nuclear Graveyard." The Bulletin of the Atomic Scientists. October 1989. 15-23.

——. "Hanford's Bitter Legacy." The Bulletin of the Atomic Scientists. January/February 1988. 17-23.

Stenehjem, Michele A. "Indecent Exposure." Natural History. September 1990. 6+.

———. "Pathways of Radioactive Contamination: Examining the History of the Hanford Nuclear Reservation." *Environmental Review*. Fall/Winter 1989. 95-112.

United States Congress, Office of Technology Assessment. *Complex Cleanup: The Environmental Legacy of Nuclear Weapons Production*, OTA-O-484. Washington, DC: U.S. Government Printing Office, February 1991.

United States Department of Energy. *Final Environmental Impact Statement: Disposal of Hanford Defense High-Level, Transuranic, and Tank Wastes*. Richland, WA, 1987.

———. Overview of the Hanford Cleanup Five-Year Plan. Richland, WA, 1991.

Washington State Department of Ecology, United States EPA, and United States DOE. *Community Relations Plan for the Hanford Federal Facility Agreement and Consent Order*. August 1989.

CHAPTER *16*

HAZARDOUS AND TOXIC SUBSTANCES: CLEAN SITES, INC.

Throughout the world, and in the United States in particular, chemicals have come to play an important role in industry, agriculture, and the lives of average citizens. Worldwide, over 70,000 different chemicals are used daily, and each year between 500 and 1,000 new synthetic compounds are introduced. Each year, the United States disposes of 270 million tons of hazardous and toxic substances—enough to fill the New Orleans Superdome 1500 times!

Hazardous and toxic substances are chemicals that can adversely affect human health and the environment. This broad definition includes chemicals such as lead, compounds such as polychlorinated biphenyls (PCBs), and the products of infectious agents like bacteria and protozoa. The term "hazardous" may conjure images of leaking, corroded barrels and abandoned, desolate waste dumps; however, these images represent only one part of the range of hazardous and toxic substances. Indeed, hazardous and toxic chemicals can be found virtually everywhere. Some are as old as the earth itself; lead and radium 222 (contained in uranium rock), for instance, are found in the earth's crust. Other substances are bound up in the planet's biota; many plants and animals, for example, manufacture substances that are poisonous, in varying degrees, to humans and other animals.

Although many toxic substances occur naturally, others are manufactured by humans through physical or chemical processes. Synthetic chemicals, many of which are highly toxic, are used in industrial processes to manufacture products—such as plastics, inks, and pharmaceuticals—for both industrial and home use. Pesticides, often used to ensure the cosmetically attractive fruits and vegetables that consumers demand, contain many potentially dangerous chemicals. The typical American home contains an amazing array of potentially harmful chemicals, from common household products (such as cleaners and shoe polish) to used motor oil, antifreeze, and insecticides.

When a toxic substance is disposed of or managed in such a way that it poses a threat to human health or the environment, it is known as a hazardous waste. Industrial and agricultural chemicals account for the vast majority of hazardous wastes. The safe disposal of these wastes, and the clean-up of hazardous waste sites, are among the most problematic of all environmental issues. Toxics can contaminate surface and ground waters, causing fish kills and destroying other aquatic life. Sensitive species are the first to be affected, and species

166

diversity in the contaminated area typically decreases. Toxics can also pollute the soil and air, sometimes dramatically upsetting the balance within the ecosystem. One of the major problems with pesticides, for example, is that they kill beneficial insects as well as pests. After the target species builds up resistance to the pesticide, its numbers can increase dramatically because the populations of its natural predators have been greatly reduced.

Despite the potential threats posed by hazardous and toxic substances, few people were aware of them throughout most of this century. In the late 1970s, however, the threat of leaking hazardous waste sites was dramatically thrust upon the public when residents of a New York community called Love Canal began to question why they had such a high number of unusual health problems ranging from rashes and headaches to cancer, respiratory ailments, miscarriages, and birth defects. Persistent efforts on the part of some citizens catalyzed government investigations, which revealed that the community's groundwater supplies, basements, storm sewers, backyards, and gardens were all seriously contaminated by toxic chemicals. Residents soon learned that their community had been built on top of a toxic waste dump.

In the aftermath of the Love Canal incident, many communities across the country learned that they, too, were at risk from improperly maintained dumps containing hazardous and toxic substances. One of the laws passed by the federal government in response to this growing realization was the Comprehensive Environmental Response, Compensation, and Liability Act, more commonly known as Superfund. Enacted in 1980, Superfund empowered the federal government to clean up hazardous waste sites and to respond to uncontrolled releases of hazardous substances into the environment.

Superfund authorizes the federal government to respond directly to releases or threatened releases of hazardous substances into the environment. The Environmental Protection Agency (EPA) manages Superfund, which is financed by taxes on the manufacture or import of crude oil and commercial chemical feedstocks (the raw chemical materials that are the first step in the manufacture of chemical and chemical-based products). If the EPA decides that a site poses significant potential harm to human or environmental health, the agency can use Superfund money to respond to the crisis in a number of ways. They may immediately remove hazardous substances from the site, conduct remedial actions such as on-site incineration of contaminated soil, water, and debris, and take measures to protect public health (including providing alternate water supplies and temporarily or permanently relocating affected residents). Whenever possible, those parties responsible for the release of hazardous wastes are forced to reimburse the fund for such expenses.

Superfund was envisioned by many to be a one-time, short-term effort at cleaning up the country's most polluted hazardous waste disposal sites. Unfortunately, the first five years of the Superfund program were marked by few victories and many setbacks. Almost half of its funds were being spent on litigation. Of the potentially thousands of dangerous waste sites in existence, the EPA was successful in cleaning up only six. By 1991, the EPA's National Priorities List (NPL) of sites requiring major cleanup actions had reached over 1200. However, only 34 sites had been cleaned up and removed from the list. Another 65 had been cleaned and were in the process of being taken off the list. Over 31,000 sites were still under investigation for possible cleanup. With an average price tag of between $21 and $30 million per site, the total cleanup cost for existing NPL sites is almost $30 billion.

In addition to affecting so few sites and costing so much, Superfund clean-ups have been criticized for lack of permanent effectiveness. In 1988, the Office of Technology Assessment issued a report evaluating the remedies selected for Superfund sites. The report found that many of the remedies used were not effective in protecting human health and the environment in the long term because they were impermanent. These measures, such as containing waste on the site, are more likely to require further action in the future than would permanent removal of wastes.

Although the Superfund process itself must be made more efficient in order to effectively deal with the staggering number of unsafe hazardous waste sites across the country, it is not the only way to deal with such sites. This chapter will focus on a private, nonprofit organization that is also helping to accelerate the clean-up of our country's hazardous waste. Known as Clean Sites, this organization was founded in 1984 by a group of environmentalists and business leaders in an attempt to increase the numbers of NPL sites being cleaned up. Clean Sites has been directly responsible for negotiating clean-up agreements among responsible parties without litigation, and it has also provided other groups with effective models for allocating responsibility, negotiating settlements, and conducting clean-ups. In addition, Clean Sites has expanded its role to include conducting research and making recommendations to improve the efficiency of Superfund and other public policies regarding hazardous waste.

PHYSICAL AND BIOLOGICAL BOUNDARIES: HAZARDOUS WASTE SITES AND THEIR EFFECTS

Most of the sites where Clean Sites has been involved are on the National Priority List. These NPL sites are usually hazardous waste storage facilities abandoned by recycling companies such as Pollution Abatement Services of Oswego, New York. Pollution Abatement Services collected hazardous waste from a variety of customers and stored it in drums at a main facility and eight satellite operations in New York state; the main facility was ranked the seventh most dangerous threat on the NPL and the most dangerous site on the New York state list.

Other abandoned recycling facilities are non-NPL, such as SED in Greensboro, North Carolina. SED collected transformers and capacitators from utility companies, drained out the polychlorinated biphenyls (PCBs), and sold the containers for scrap. SED left the site with six million pounds of PCBs stored in two rented warehouses and eleven trailers. Because SED abandoned this site after Congress passed the Superfund legislation and established the NPL, it was not included on the list despite its considerable threat to the environment. Other sites include sanitary landfills, intended only for non-hazardous wastes, where hazardous substances were carelessly or unknowingly deposited.

Dangerous chemicals contaminating these sites include dioxin, furan, PCBs, phenols, chloride, boron, and ammonia. At most sites, operators did not properly dispose of the waste that contains these substances, and so it is able to leak into the groundwater. Groundwater supplies 25 percent of the water we use in our households and in industry, and toxic materials entering groundwater can easily affect the population surrounding the disposal site. By entering major rivers and lakes, groundwater can spread hazardous material farther. In some cases, toxic substances even leak directly into the surface water.

SOCIAL BOUNDARIES: THE MANY OBSTACLES
BLOCKING HAZARDOUS WASTE CLEAN-UP

Through much of the history of hazardous waste disposal, industrial development, especially in the chemical industry, has taken priority over possible environmental hazards. However, legislation like Superfund and companies like Clean Sites are part of a growing attempt to bring the two considerations into balance by making those who produce toxic waste responsible for the environmental consequences.

One major reason why responsible parties failed to clean up NPL sites after the passage of Superfund was economic confusion: what parties were responsible for paying for a clean-up, and how much did they owe? Before Superfund, companies would send hazardous waste to a recycler or a landfill and never expect to think about it again. However, Superfund made the producers of hazardous waste responsible for all of its environmental impacts, even after they had passed it on to someone else. As a result, many companies were shocked and angry when they found they had to pay for expensive clean-ups of waste they thought was out of their hands forever. Many attempted to shift the blame to other parties or simply avoided paying. These sorts of problems contributed to Superfund's widely recognized lack of efficiency, and prompted the founding of Clean Sites.

LOOKING BACK: THE HISTORY OF CLEAN SITES

Clean Sites' conception might be traced back to August 8, 1982, when an article by Christopher Palmer of the National Audubon Society appeared in the *Washington Post*. He called for business leaders and environmentalists to work together to preserve the land, air, and water of their environment. In response, Louis Fernandez, chairman of Monsanto Company, formed a steering committee to explore the problem of hazardous waste clean-up, an issue of special interest to chemical companies that produce a great deal of hazardous waste; the committee included representatives from the Conservation Foundation, the National Wildlife Federation, Exxon Chemical Co., the Chemical Manufacturers Association, and E.I. DuPont de Nemours & Co.

Between 1982 and 1984, this diverse group worked on the problems of hazardous waste clean-up. Given the failure of responsible parties to respond to Superfund, the committee concluded that the situation required more attention than the EPA could give it. They proposed forming a third-party organization that could act as a catalyst to encourage clean-up at inactive dump sites, especially those where potentially responsible parties could be identified.

How Clean Sites Began

On May 31, 1984, the resulting organization was introduced to the public as Clean Sites, Incorporated. Clean Sites officials stressed that the group would supplement, not substitute for, the efforts of the EPA and Superfund. With the support of EPA Administrator William Ruckelshaus, Clean Sites leaders hoped to facilitate clean-up of NPL sites by mediating negotiations among potentially responsible parties and between those parties and government agencies. Clean Sites' goal was to begin working toward clean-ups at 20 sites in the first year. Since clean-up of the thousands of dangerous hazardous waste sites in the

United States would require the efforts of many groups besides Clean Sites, they also resolved to provide those groups (including environmental consulting firms and responsible parties that undertake clean-ups on their own) with tested and proven models of how to allocate responsibility among potentially responsible parties, how to negotiate settlements, and how to conduct remedial clean-up activities.

Clean Sites almost lost the chance to realize its goals because of a problem few had anticipated: in the summer of 1984, the private environment liability insurance market dried up, leaving Clean Sites unable to find insurance coverage despite a world-wide search. If Clean Sites became active, its directors and officers would face personal liability in case of a lawsuit stemming from the group's actions as a facilitator or advisor.

At that point, Charles Powers, president of Clean Sites, told the EPA that he gave the organization a life expectancy of one month. However, in February 1985, the EPA agreed to cover Clean Sites' liability for up to $5 million per site and a total of $10 million per year with money from the Superfund trust fund. This indemnification covered Clean Sites' negotiating and consulting services; in cases where they actually worked on a site, such as project management, the responsible parties had to provide insurance for those activities. This agreement required that Clean Sites seek EPA approval for each site before beginning any sort of negotiations or work and that Clean Sites continue to search for another source of insurance. Also, the agreement prevented Clean Sites from accepting payment from responsible parties for settlement, technical, or management services that it provided at sites covered by the EPA.

How Clean Sites Funds Activities and Clean-ups

Although Clean Sites' agreement with the EPA prevented it from charging for its services, Clean Sites also received no government money for operating costs because it was a private corporation. These costs, which Clean Sites estimated would be $5 million in 1984 and between $12 and $15 million for the next few years after that, had to be funded by donations. In an exemplary move toward environmental responsibility, members of the Chemical Manufacturers Association agreed to provide up to one half of Clean Sites' operating funds for its first three years. Clean Sites hoped to obtain the other half from foundations and other industries involved with hazardous waste, including petroleum, paper products, steel, nonferrous metals, electronics, rubber, and health care companies. Although Clean Sites received $2.3 million from approximately 100 chemical manufacturers in its first year, contributions from other sources proved well below expectations. Fortunately, the Hewlett and Mellon Foundations made grants to Clean Sites for a total of over $1.3 million between 1986 and 1988.

Although Clean Sites was founded as a strictly nonpartisan group, some environmentalists were worried that its large subsidies from the chemical industry would affect the group's ability to be objective. However, careful selection of the Board of Directors to include officials from major environmental groups like the National Wildlife Federation and Audubon Society helped guard against any conflict of interest. Eventually, though, in 1986, Clean Sites received permission from the EPA to begin charging responsible parties for its services, further supplementing its budget. By 1989, Clean Sites earned over half its income from payments for services, and received less than 33 percent of it funds from chemical companies.

The clean-ups themselves have always been funded by the responsible parties. When Clean

Sites began, each clean-up had to be totally funded by responsible parties because the EPA could not make Superfund money available for costs that remained unallocated. In cases like the Bayou Sorrel site in Louisiana, one of Clean Sites' tasks was to find additional potentially responsible parties to cover costs that the parties involved in negotiations would not agree to pay. However, in 1986, the Superfund legislation was reauthorized, and the new version allowed the government to provide some money for mixed funding sites, where cleanups are funded both by responsible parties and Superfund. Clean Sites participated in two of the first three mixed funding settlements after the 1986 reauthorization.

How Clean Sites Operates

Although Clean Sites never actually cleans up a site, it provides a wide variety of services including settlement, technical review, and project management. Before Clean Sites provides any of these services, it must first decide on an appropriate site. Clean Sites' activities concentrate on NPL sites, where it usually finds potentially responsible parties and informs them of their responsibility for the clean-up. Clean Sites also considers other sites based on suggestions from private citizens, potentially responsible parties, environmental groups, community organizations, and government agencies. Clean Sites officials choose sites where they think no action will be taken without their involvement.

One of Clean Sites' most important and sought-after services is settlement, which primarily involves negotiating clean-up agreements among potentially responsible parties. The first step involves finding potentially responsible parties (although occasionally such parties contact Clean Sites) and allocating responsibility for the hazardous waste. For instance, at a garbage dump in Ripon, Wisconsin, Clean Sites helped the owners and users reconstruct how much hazardous material they had each dumped over the years. Next, Clean Sites must help all the responsible parties agree on a payment plan. Because sites can involve hundreds of parties (the Rose Chemical site in Missouri involved 750), this process might be almost impossible without a nonpartisan mediator like Clean Sites.

In addition to mediating agreements, Clean Sites can provide nonbinding arbitration, a service for which demand is growing. Rather than simply helping parties develop and agree on a course of action, Clean Sites makes the first attempt at dividing clean-up costs among responsible parties. Then the parties decide whether or not to accept the plan.

After mediation or arbitration, responsible parties at some sites will then request the assistance of Clean Sites' technical review service to help develop a site-specific program for clean-up. In most cases, responsible parties hire contractors to conduct two major studies of a site: a remedial investigation to determine contamination levels and pathways and a feasibility study to evaluate the different ways to clean up the site. Clean Sites examines the studies to make sure they meet government requirements and then passes them along to the EPA, which approves a particular plan for clean-up. At times, waste simply needs to be better contained by methods like building underground walls to prevent movement into groundwater. Frequently, though, waste must be incinerated or removed to properly maintained secure landfills, which are designed to safely hold hazardous materials. Sometimes Clean Sites even oversees the next step: designing the plan to carry out the clean-up.

At some sites, especially those involving many responsible parties, Clean Sites is hired to manage the actual clean-up of a site. In these cases, the project management service coordinates the use of funds and organizes the operations of the clean-up. First they estimate costs for each task involved and then schedule those tasks. Then, they select and monitor contractors who carry out the clean-up.

Clean Sites can fulfill other functions at a site as well. They often mediate negotiations for clean-up plans between potentially responsible parties and the government. Another important function they can perform is to keep the communities near a site informed about progress in clean-up activities. An article in the local newspaper of Holden, Kansas—the community near the Rose Chemical clean-up— expressed how much the residents and local businesses appreciated Clean Sites' efforts to keep them up-to-date. Clean Sites has also assisted state government and federal agencies in preparing methods for hazardous waste assessment and clean-up, and has participated in educational programs such as teaching citizens how to apply for EPA grants and training workers how to handle hazardous waste.

How Clean Sites Has Changed over the Years

Over the few years since its founding, Clean Sites has progressed both in the number of sites where it has been active and in the variety of services it provides. By the end of its first year, it was involved at 19 sites, one fewer than initially projected. However, two were cluster sites involving many sites closely situated within the same geographic area, which brought the total up to 45. In fact, Clean Sites officials felt the cluster approach was a major step toward improving the cost-effectiveness of clean-ups. By sharing expensive procedures like importing or building an incinerator, each site can substantially reduce costs.

The number of sites where Clean Sites was active continued to grow, reaching 71 by 1991. Although Clean Sites fell short of its initial goal of 60 sites per year, the organization is generally considered a success. One major achievement is that potentially responsible parties are beginning to realize the advantages of coming to an agreement independently or through a group like Clean Sites. In that way, they can gain some control over management costs, technical studies, and the method of clean-up. In fact, Clean Sites estimates that, through their involvement, they save responsible parties 70 percent of clean-up costs through the normal Superfund process, where the government cleans a site and then sues responsible parties for the costs.

Although thousands of NPL sites still remain polluted, and tens of thousands more may need attention in the future, Clean Sites has played an important role in dealing with the problem. Besides accelerating clean-up at specific sites, it has provided effective models for others to use, such as the 1987 booklet, *Allocation of Superfund Site Costs through Mediation*. Thomas Grumbly, president of Clean Sites, feels that providing information so others can run effective clean-ups, too, is one of the best ways Clean Sites can combat the overwhelming number of untouched hazardous waste sites, and he anticipates that Clean Sites will expand this role in the future. He compares Clean Sites' role to that of a research hospital that not only helps specific patients, or sites, but also learns more about the processes involved so it can pass the information on to policymakers and other groups involved in clean-ups.

How Clean Sites Shapes Public Policy

Since Grumbly took office in 1987, Clean Sites has become increasingly more involved in shaping public policy affecting hazardous waste clean-up. In 1989, Clean Sites offered a memo to President Bush entitled *Making Superfund Work*. This memo explained how the EPA can manage cleanup of NPL sites more efficiently by implementing

- stronger efforts to make responsible parties pay for their shares of EPA-initiated cleanups at NPL sites—preventing the EPA from using up Superfund money on clean-ups that others should be funding.
- a better guide to Superfund's complex, confusing clean-up standards.
- more efficient management, giving regional EPA officials more decision-making authority.
- new definitions of clean-up success, such as speed, quality, and percentage financed by responsible parties.

This unprecedented criticism of a government agency signalled a new stage in Clean Sites' growth into a major force in hazardous waste clean-up in the United States

In 1990, Clean Sites' public policy efforts extended to improving the Superfund remedy selection process. Using funds from a $135,000 grant from the EPA and a matching amount from the Andrew W. Mellon Foundation, Clean Sites conducted a comprehensive study that brought together over 100 experts from government, industry, and public interest groups. The report resulting from the study, "Improving Remedy Selection: An Explicit and Interactive Process for the Superfund Program," cited two major problems with the EPA's method of remedy selection. The first was the EPA's policy of spending a great deal of time and money to explore numerous clean-up methods before deciding on the level of protection required by a specific site. The report recommended that the EPA instead develop explicit protection objectives for each site early on; factors used to determine such objectives should include the risks posed by the site, the site's future use, and the concerns of nearby residents. Then, the EPA should explore only those remedies that will meet the predetermined objectives. The second problem the report cited was the need for the EPA to better define general goals and objectives for clean-ups, specifically in regard to permanence and long-term effectiveness.

Clean Sites also contributes to public policy by providing research information. One such project, funded by a grant from the Ford Foundation, recently examined how hazardous waste sites in poor, rural areas affected groundwater; because 97 percent of America's rural poor depend on groundwater for drinking, contamination by toxic waste could dramatically affect their health. The study found that there was significant contamination of groundwater at many sites. Although at some sites, plans already existed to clean up the waste or to provide alternative sources of drinking water, Clean Sites also determined that these poor, rural sites were less likely than others to be placed on the NPL.

Another important way Clean Sites has expanded its efforts to assist other groups in cleaning up hazardous waste sites affects state Superfund programs. These programs target some of the thousands of non-NPL sites that nevertheless pose significant threats to human health and the environment. In 1991, Clean Sites joined forces with the Environmental Law Institute to form the State Superfund Network. This network enables state programs to exchange information, thus avoiding costly mistakes and increasing efficiency in areas such as funding, remedy selection, enforcement, and settlement negotiation.

LOOKING AHEAD: THE FUTURE OF CLEAN SITES AND HAZARDOUS WASTE CLEAN-UP

In order to continue its successful role in our country's efforts to clean up hazardous waste sites, Clean Sites must be careful to maintain its objectivity and nonpartisan status, invaluable credentials that convince both responsible parties and the government to accept it as a mediator. To preserve this status, Clean Sites must continue to maintain a balance of environmental organizations and private industry on its board and among its supporters.

However, despite Clean Sites' growing reputation as a mediator and arbitrator of clean-up agreements, it can directly intervene in only a fraction of the unsafe hazardous waste sites around the country. Therefore, it should persist in expanding its efforts to provide other groups with models for dealing with all aspects of hazardous waste clean-ups. In addition, Clean Sites should continue its growing involvement in shaping public policy in general and in helping the EPA more efficiently manage NPL sites in particular.

SUGGESTED READING

Clean Sites. *Annual Report*, 1989 and 1990.

"Clean Sites at Age One . . . An Uneven Start." *The Environmental Forum*. July 1985. 27-31.

Garon, Stephen, and Melinda J. Holland. "Dispute Resolution Through Mediation." *National Environmental Enforcement Journal*. October 1991. 3-10.

Green, Rick. "Year of the Superfund." *Sierra*. May/June 1985. 36+.

Grumbly, Thomas. "Making Hazardous Waste Cleanup and Science Compatible." *Environmental Science and Technology*. July 1991. 1185.

——. "Superfund: Candidly Speaking." *EPA Journal*. July/August 1991. 19-22.

Hanson, David J. "Clean Sites, Inc., Moves into Full Swing with Waste Cleanup." *Chemical and Engineering News*. October 26. 1987. 14-16.

Holland, Melinda J., and Stephen Garon. "Clean Sites' State Superfund Assistance Activities." *National Environmental Enforcement Journal*. September 1991. 3-7.

Improving Remedy Selection: An Explicit and Interactive Process for the Superfund Program. Alexandria, VA: Clean Sites, October 1990.

Koshland, Daniel E. "Toxic Chemicals and Toxic Laws." *Science*. August 30 1991. 949.

Lappe, Marc. *Chemical Deception: The Toxic Threat to Health and the Environment.* San Francisco: Sierra Club Books, 1991.

Long, Janice. "President Urged to Act on Waste Clean-up." *Chemical and Engineering News.* January 30 1989. 22-3.

MacKerron, Conrad. "Clean Sites Hits its Stride as a Super Mediator." *Chemical Week.* April 27 1988. 19-20.

——. "Memo to Bush: How to Make Superfund Work." *Chemical Week.* February 1989. 20.

Magnunson, Ed. "A Problem that Cannot Be Buried." *Time*. October 14 1985. 76-84.

Making Superfund Work: Recommendations to Improve Program Implementation. Alexandria, VA: Clean Sites, January 1989.

Rich, Laurie A. "Clean Sites, Inc.: Sweeping Aside Startup Problems and Getting on with the Job." *Chemical Week*. June 5 1985. 28-31.

Rikleen, Lauren Stiller. "Superfund Settlements: Key to Accelerated Waste Cleanups." *The Environmental Forum*. August 1985. 51-54.

SanGeorge, Robert. "Clean Sites, Inc." *Environment*. December 1985. 5+.

"Speeding Hazardous Waste Site Cleanup: Industry, Conservationists Work Together." *ChemEcology*. May 1984.

Stutz, Bruce D. "Cleaning Up." *The Atlantic*. October 1990. 46+.

Thies, Austin C. "Successful Cleanup through Effective Cooperation." *Environment*. April 1986. 32-34.

Travis, Curtis C., and Carolyn B. Doty. "Superfund: A Program without Priorities." *Environmental Science and Technology*. November 1989. 1333-1334.

U.S. Congress, Office of Technology Assessment. *Are We Cleaning Up? 10 Superfund Case Studies—Special Report*. OTA-ITE-362. Washington, DC: U.S. Government Printing Office, June 1988.

Wolf, Douglas W. "Superfund: The Polluter Must Be Made to Pay." *Environment*. January/February 1989. 42-44.

Wolf, Frederick. "Superfund: An Alternative Funding Approach." *Journal of Environmental Health*. January/February 1992. 18-22.

CHAPTER 17

UNREALIZED RESOURCES—WASTE MINIMIZATION AND RESOURCE RECOVERY: THE NEW ALCHEMY INSTITUTE

Until recently, few people thought about what would happen to their old food containers, newspapers, lawn clippings and other waste after they threw them away. Today, however, the general public is coming to recognize that the disposal of our waste can be a serious environmental problem. Because the human population is growing, the sheer volume of wastes we produce is on the rise. In the United States, waste disposal is an especially pressing issue because, as citizens of an industrialized nation, we tend to have higher standards of living and consume more goods and products per capita than people in less developed countries. In addition, because of the nature of our technological and consumption-oriented culture, we throw away a great deal of elaborate packaging and synthetic products. In less developed countries and even other industrialized nations, people are more likely to use materials and products efficiently and to re-use or recycle them.

In addition to producing a growing amount of garbage, we are running out places to put it. Landfills, where we have buried much of our refuse in the past, are rapidly filling. Many communities are questioning the wisdom of developing new landfills because of the potential risks to human health and the environment. Ocean dumping is also on the decline because of its adverse affect on water quality and marine life. Incineration, another option for waste disposal, can be very costly and may pose its own environmental and health hazards (see Chapter 10).

In the face of this escalating "garbage crisis," it is becoming clear that we must search for other options that are sustainable over time—that won't run out of room or harm the air or water we depend on. Perhaps the most sound strategy is the one we're hearing more often from environmentalists and city planners alike: "reduce, re-use, recycle." Indeed, a great deal of our waste, such as excessive packaging, is unnecessary. In addition, a surprising amount of what we discard could be considered an unrealized resource that could be used to benefit both human and natural systems. Whether the resource is municipal sewage wastes, hazardous wastes, or nonhazardous solid wastes, reusing and recovering these substances will enable human systems to more closely mirror natural systems and lessen the impact of human activities on the environment.

Most of us are familiar with the idea of recovering "unrealized" resources by recycling glass, paper, and aluminum. However, we often overlook another valuable resource disguised as waste: organic material, such as grass clippings, animal manure, or food scraps. This material contains carbon, the essential ingredient of life. Organic material can be recycled even more easily than other wastes because it decays naturally in compost piles, eventually producing a rich fertilizer. In this chapter, we will focus on New Alchemy Institute (NAI), a nonprofit research and education center that specializes in developing innovative methods of nutrient recycling and composting (as well as other economical and ecologically sound ways for individuals to become more self-sufficient). Like the medieval alchemists who sought to transform worthless metals into gold, the New Alchemists seek to transform material normally discarded or ignored into resources for producing food, energy, and shelter.

The New Alchemy Institute has designed simple and affordable "technologies," including a composting greenhouse, which make reducing and recycling human, animal, and vegetable waste a realizable goal for anyone who is interested. Mimicking the cycles of nature, they use cheap, easy to obtain materials: living bacteria and plants for machinery and organic wastes and the sun for energy sources. In addition to using organic resources more efficiently, composting also reduces the amount of waste that must be transported off the site and produces a rich fertilizer that can be used or sold. This organically produced fertilizer can help farmers decrease reliance on chemical fertilizers, which are made from fossil fuels.

PHYSICAL BOUNDARIES: THE COMPOSTING GREENHOUSE AND OTHER NEW ALCHEMY INSTITUTE FACILITIES

The New Alchemy Institute is located near Cape Cod in East Falmouth, Massachusetts, on the 12-acre site of a former dairy farm. The grounds encompass a variety of experiments and demonstrations—including but not confined to those that recycle organic material as compost—that promote self sufficiency and ecological awareness. These energy and cost-efficient projects are designed to meet the needs of homeowners, small-scale farmers, and educational institutions.

Many visitors enjoy strolling through the New Alchemy Institute's organic gardens, which are cultivated with compost and manure rather than synthetic fertilizers, herbicides, or pesticides. The gardens feature vegetables grown for market as well as flowers and herbs. Another enjoyable and educational site is the Visitors' Center and Store, which houses exhibits and a slide show describing NAI's history and ongoing projects. The store also offers books, tools, and garden supplies for sale. Another educational facility is the Superinsulated Auditorium, a converted dairy barn that demonstrates advanced techniques for conserving energy. These techniques include superinsulation, air-vapor barriers, an air-to-air heat exchanger, low-flow toilets, and a tankless hot water heater. Boasting three times the insulation and one-tenth the air leakage of a conventional structure, the 1400 foot square Auditorium costs only $150-250 to heat annually.

NAI's facilities also include a variety of greenhouses. The Cape Cod Ark was built in 1976 as NAI's first bioshelter, or solar greenhouse designed to sustain fish, plants, and other life forms with no energy except sunlight. The Pillow Dome is a geodesic dome 30 feet in diameter. An aluminum frame supports triple-layered triangular plastic pillows that are inflated with inert gas—"super" insulation that helps to retain heat in the wintertime. Inside, vegetables grow and solar aquaculture ponds sustain fish and store heat. The Composting Greenhouse, which recycles organic materials, is an especially good example of resource recovery, and thus we will explore it in more depth throughout this chapter.

The Theory Behind the Composting Greenhouse

Researchers at the Institute became interested in developing a sustainable, affordable greenhouse in the early 1980s. Traditional solar greenhouses are very expensive to build because they are designed for use without supplemental heat. Consequently, they must be heavily insulated to withstand the winter months, and insulation blocks sunlight and adds to cost. Various storage systems that "save" heat from the sun are an additional cost.

In 1983, NAI built the composting greenhouse with the help of the Biothermal Energy Center. This greenhouse was an attempt to eliminate insulation and heat-storage costs by taking advantage of the heat generated when bacteria decompose organic material in compost. Using the heat from compost to supplement the sun's rays not only cuts costs without relying on nonrenewable energy, but also promotes better management of manure, which can easily pollute water if improperly stored. Finally, the carbon dioxide, ammonia, and moisture produced along with the heat all benefit the plants as well.

How the Composting Greenhouse Works

The basic structure of the greenhouse is a 12 by 48 foot frame (576 square feet) fashioned from electrical conduit. A double layered plastic film, inflated to provide extra insulation, makes up the shell of the greenhouse. The greenhouse is situated on an east-west axis, with the 25 cubic yard composting chamber located on the north side of the greenhouse. The compost chamber has been conveniently designed with removeable, insulated panels so it can be loaded from the outside.

The upper level growing beds are located on top of the compost chamber, and the lower level beds are in front of the chamber. Blowers periodically pull air from the greenhouse underneath and up through the compost. This ventilation system eliminates the labor- and energy-intensive practice of turning the compost every other day in order to provide oxygen to the aerobic bacteria that decompose the organic waste. Blowing through the compost, the air picks up heat, moisture, carbon dioxide, and ammonia produced by the bacteria and other organisms and carries it into perforated pipes running beneath the lower growing beds.

The water molecules in the compost store energy; when the moisture condenses in the pipes, it releases the stored energy into the soil in the form of heat. The water is also absorbed by the soil and, eventually, by the plants. (Although the compost is capable of providing all of the necessary moisture required by plants and bacteria, a backup watering system is available just in case.) The carbon dioxide that enters the pipes filters up through the soil and becomes available to the underside of the leaves of the plants.

Unlike the other compost by-products, ammonia can be damaging to plants even in low concentrations. This potential danger is transformed into a benefit by the soil in the growing beds, which acts as a biofilter: the bacteria living in the soil transform the damaging ammonia into useful nitrate, which is readily absorbed by plant root hairs. The texture and health of the biofilter soil are enhanced by populations of earthworms and manure worms that create vertical tunnels through the soil as they digest organic matter present in the topsoil.

Although the NAI composting greenhouse uses soil and plants as a biofilter, many composting sites use other media. At some sites, odor control is especially important (for instance, if sewage sludge is being composted near populated areas). In these cases, screened loam, peat, and woodchips are effective in eliminating ammonia, mercaptans, hydrogen sulfide, ethyl acetate, and other smelly gases.

The compost itself can be composed of a variety of organic materials, including sewage sludge, human waste (nightsoil), woodchips and sawdust, animal manures, garbage and yard wastes, garden and vegetable wastes, and even preshredded municipal solid waste mixed with sludge. In each case, the materials decompose at different rates and give off varying amounts of carbon dioxide and heat. Anyone operating a composting greenhouse must evaluate individual plant needs and seasonal differences to determine what material will work best. In order to produce the appropriate amounts of heat and carbon dioxide as well as a high-quality compost, greenhouse owners must often mix several materials.

BIOLOGICAL BOUNDARIES: NATURAL CYCLES OF THE COMPOSTING GREENHOUSE

By setting up natural cycles of decomposition and growth within the composting greenhouse, NAI has created a self sufficient system. All NAI workers must do is replace the ready, finished compost with new compost material, fertilize or replace the soil with finished compost, and care for the plants.

Sunlight provides the basic energy the plants need to photosynthesize and grow. Naturally occurring biological organisms perform the alchemy of turning the organic wastes into nutrient-rich soil that can be used as fertilizer. These microbes are aerobic; that is, they need oxygen to live. In addition, they use carbon as their "fuel." In the natural environment, plants give off oxygen. When plants and other living things die, they release carbon into the soil. Mimicking these natural cycles, the composting greenhouse provides the microbes with oxygen from the plants growing in the greenhouse and with carbon from the organic material in the compost. Taking in oxygen and carbon, they give off moisture, heat, carbon dioxide, and ammonia as waste products. To plants, however, these wastes are essential to life. The NAI composting greenhouse uses these products to grow healthy plants in addition to producing fertilizer.

Heat and moisture, which are bound up together, are required for the plants' basic metabolic processes. Because the sun can produce sufficient heat for the greenhouse in the warmer months, heat from the compost is needed only during cold weather. However, the other by-products of composting are used to supplement plant growth all year long. Carbon dioxide is essential for the process of photosynthesis, by which plants produce their own "food." In fact, NAI researchers eventually realized that the extra carbon dioxide from the compost was even more valuable than the heat. The added carbon dioxide in the greenhouse has been shown to produce 20-30 percent net increases in yields of vegetable crops. Nitrogen is a limiting nutrient in most greenhouses, and a short supply of it retards the growth of plants. Thus, the increased nitrogen in the greenhouse also helps plant growth significantly.

Overall, the self-sustaining nature greenhouse has proven very successful: it produced over 100 tons of compost and tens of thousands of seedlings in its first full year. NAI researchers estimate that for every cubic yard of compost materials they place in the composting chamber, they save approximately $5.50-11.00 they would have had to spend on supplemental CO_2 and $3-6 that would have gone for heat.

SOCIAL BOUNDARIES: THE MISSION OF THE NEW ALCHEMY INSTITUTE

New Alchemy is a philosophy and a proposed way of life as well as a research institute. The ideas that lie behind the Institute grew out of concern over the increasing technologization and specialization of our world. Not only is our reliance on complex, energy-intensive technology destroying the environment, but it is also eliminating the natural diversity of plants and animals, leaving us vulnerable if a species we rely upon succumbs to disease or predators. Similar species that could have served the same function may well have been considered "useless" and driven to extinction. In addition, this reliance on technology has led to a society of individuals with highly specialized skills. Because so many of us have nothing to do with producing our own food, shelter, or energy, we know little about how to provide these things for ourselves if we need to. This ignorance could be fatal if the technology and mass production that supply our needs fail us someday.

The research at NAI is meant to address these concerns by enabling people to coexist in a mutually supportive and beneficial way with nature. The NAI philosophy holds that the environmental and social problems we have created through complex, energy-intensive technology cannot be solved by more of this type of technology; thus, NAI concentrates on providing biological analogues to chemical and mechanical methods of producing food, shelter, and energy.

The NAI philosophy also includes the belief that we must change as a society in order to achieve self-reliant, ecologically harmonious lifestyles. All of NAI's projects work toward enabling individuals to fulfill more of their own basic needs without overly complex, energy-intensive technology or dependence on centralized services. For instance, the composting greenhouse is a simple, affordable way for any individual to produce food and the compost that will keep his or her soil fertile while simultaneously disposing of much of the organic waste he or she produces. The ultimate vision of the NAI is a sustainable society, composed of self-reliant communities. These communities would grow all of their own food and generate their own power from renewable resources. On a smaller scale, their main focus has been to create a small, self-reliant, organic farm.

LOOKING BACK: THE HISTORY OF THE NEW ALCHEMY INSTITUTE

Since its founding almost 20 years ago, NAI has been on the forefront of a new, simpler, less energy-intensive technology: appropriate technology that is life-encouraging and ultimately safer and simpler than its chemical and mechanical analogues. By developing naturally based methods to increase self-reliance and educating the public, they are trying to provide answers today to the problems of tomorrow.

Origins of the New Alchemy Institute

The founders of the New Alchemy Institute, John and Nancy Jack Todd and Bill McLarney, originally lived in southern California. The idea for NAI evolved in the late 1960s, when John was teaching biology classes at San Diego State. Ecology and environmental degradation were just beginning to gain widespread public awareness, so John and Bill began to hold informal seminars on certain aspects of the overall problem. These seminars often lasted late into the night and focused on the question "Can anything be done?" and if so, "What?"

John had also been exploring the practical side of ecological awareness by taking his students on field trips to the hills east of San Diego. His original purpose was to examine the area and discuss how to farm the land without degrading it. However, John soon realized that most students had no sense of the land and how to use its resources. A second phase of the project began in which the students learned to group species of plants and animals taxonomically as well as to classify soil types. By examining each species in its environment, they began to recognize certain patterns in the land. For instance, the existence of a plant that only grows near water in a seemingly dry area provides a clue that there is probably a hidden spring nearby. Such an ability to "read" a landscape is essential for those who want to farm in harmony with the land with a minimum of dependence on technological devices.

These experiences inspired and initiated the philosophy that drives New Alchemy today. In 1970, John Todd expressed the essence of the New Alchemy ideal in a paper titled *A Modest Proposal*. This essay served as the first New Alchemy newsletter and has since been translated into many languages. In *A Modest Proposal*, John sets the ground for a new vision of a biologically harmonious plan for the future of all people. He believed that

> . . . a plan for the future should create alternatives and help counter the trend toward uniformity. It should provide immediately applicable solutions for small farmers, homesteaders, native peoples everywhere, and the young seeking ecologically sane lives, enabling them to extend their uniqueness and vitality. Our ideas could also have a beneficial impact on a wider scale if some of the concepts were incorporated into society-at-large. Perhaps they could save millions of lives during crisis periods in the highly developed states.

Research and Education at New Alchemy Institute

As the first New Alchemists began developing their ideas into research projects, the group looked for funding and a site to conduct the research. Once they found the site on Cape Cod, they were settled in by May 1972. Research projects concentrated on human support systems essential for human survival, including wind energy, solar energy, and organic and greenhouse agriculture.

The creation of the Cape Cod Ark bioshelter was the cornerstone of the early work at NAI. This solar greenhouse meets the energy, food, and shelter demands not only of humans, but also of bacteria and soil, plants, trees, flowers, insects, and fish. It is a semi-contained ecosystem that conserves energy and recycles all materials. This innovative use of natural cycles and processes to meet human needs led to the diverse projects going on at NAI today.

In addition to designing systems for meeting basic human needs, NAI has also established a research station in Costa Rica. This station concentrates specifically on investigating solutions to the destruction caused by deforestation and offers the country help in avoiding the environmentally and economically disastrous mistakes that many developing nations make in their quest to modernize.

Currently, NAI is working with twenty New England farmers to test integrated pest management and cover cropping systems. In integrated pest management, farmers use biological controls in place of chemical pesticides. Growing cover crops on land that would normally be bare over the winter helps to sustain soil fertility, control weeds, and reduce erosion without the use of chemicals.

Besides researching and developing various biotechnical devices and methods, New Alchemy staff concentrate on educating people about other aspects of an ecologically sound, self-reliant lifestyle. They offer a catalogue of publications and devices that can assist individuals in achieving this way of life. Their quarterly newsletter reports on NAI research and activities, along with updates on advances in small-scale, low energy input technology made by other organizations or individuals. NAI staff give tours, workshops, and programs for local schoolchildren, and courses and lectures for adults. In addition, the Institute offers internships for college students. They also produce technical reports evaluating various ecologically sound methods of producing food, energy, and shelter, such as non-toxic pest management or solar greenhouses.

New Alchemists Move On

In 1980, John and Nancy Todd left New Alchemy Institute to form two new corporations: Ocean Arks International, a nonprofit research organization, and The Four Elements Corporation, a for-profit enterprise set up to prove that the Todds' ideas were economically feasible. One of Ocean Arks' first projects was to develop a sailing fishing vessel for use in underdeveloped countries to help those who fish to reduce their dependency on fossil fuels. The boats were even designed so that fishers could grow the materials to make them!

Ocean Arks' most exciting design has been a solar aquatic sewage treatment facility. This facility treats sewage by cycling it through a greenhouse containing a series of tanks housing different organisms and an artificial marsh. In this mini-ecosystem, mineral, chemical, biological, and gas cycles all function together, as they do in nature, to use organic waste as food for new life: flowers, vegetables, fish, and other organisms. In the process, water is purified without the use of chemicals. Although this biological system would not be economically feasible for individuals, it makes a perfect centralized treatment system for communities that want to find ecologically sound alternatives to chemical and mechanical processes.

LOOKING AHEAD: THE FUTURE OF THE NEW ALCHEMY INSTITUTE

New Alchemy Institute has been a leader in establishing nonharmful technologies and biological analogues to chemical and mechanical technologies. Through efforts like the composting greenhouse, they have demonstrated that it is possible to work with sunlight and life, as nature does, to recycle our waste efficiently and safely (for ourselves and the environment).

Education is one of the most important functions that will help the New Alchemy Institute achieve its goals— their small-scale, low energy input technology will only make a difference if it is adopted by individuals and communities. By spreading their ideas and products among as many people as possible, NAI could set the stage for a widespread movement toward environmentally sound ways to provide food, shelter, and energy.

SUGGESTED READING

Burke, William K. "Restoring Water Naturally." *Technology Review*. January 1991. 16-17.

Hallowell, Christopher. "Plants that Purify." *Audubon*. January/February 1992. 76-80.

Jereski, Laura. "This Greenhouse Effect Just Might Be Good For Us." *Business Week*, September 18 1989. 119-120.

Luoma, John. "Trash Can Realities." *Audubon*. March 1990. 86-97.

New Alchemy Institute (237 Hatchville Road, East Falmouth, Massachusetts, 02536). *A Visitor's Guide to New Alchemy's Composting Greenhouse*.

——. *A Visitor's Guide to the New Alchemy Gardens*.

——. *New Alchemy Quarterly*. Published since 1981.

——. *The New Alchemy Institute Catalogue: Books and Products for Ecological Living*. 1989-1990.

——. *Research Report No. 3: The Composting Greenhouse at NAI: A Report on Two Years of Operation and Monitoring*.

"Reducing the Garbage Glut: Follow the Three R's - Reduce, Reuse, and Recycle." *International Wildlife*. May/June 1990. 25.

Rogoff, Martin H., and Stephen L. Rawlins. "Food Security: A Technological Alternative; Biotechnology Can Convert Biomass into a Stable Food Supply." *Bioscience*. December 1987. 800-807.

Sides, Susan. "Compost: How to Make It and How to Use It: 7 Tips for Success." *The Mother Earth News*. August/September 1991. 48-52.

Todd, Nancy Jack, ed. *The Book of the New Alchemists*. New York: Dutton, 1977.

——, and John Todd. *Bioshelters, Ocean Arks, City Farming: Ecology as the Basis of Design*. San Francisco: Sierra Club Books, 1984.

Weintraub, Pamela. "Interview with John Todd." *Omni*. August 1984. 76+.

CHAPTER *18*

PUBLIC LANDS: YOSEMITE NATIONAL PARK

When settlers first began to expand across the North American continent, they found what seemed to be endless expanses of prairies, forests, and mountains. However, by the late 1800s, some Americans realized that natural areas would have to be protected against the wave of development, farming, logging, and other threats that had swept the country. It was this realization that eventually gave birth to our country's system of National Parks.

Some historians suggest that Americans' early desire to preserve natural areas in an unspoiled state stemmed from a national sense of inferiority about the great cathedrals, castles, and ruins of European cultures. Lacking such historic resources, Americans instead valued the monuments of nature. Thus, the first preservation efforts were aimed at areas that boasted unusual beauty or spectacular features. One of those that were unique was the area now known as Yellowstone National Park.

Throughout the 1800s, fantastic tales of the phenomena to be found in Yellowstone spread around the country. Finally in 1870, a party of explorers confirmed the stories of boiling springs and towering geysers of water, although they discredited the descriptions of mountains of glass and petrified birds. Photos and paintings of Yellowstone circulated around the country, fueling popular sentiment for protecting the area from exploitation. People felt strongly that it should not go the way of Niagara Falls, once considered one of the country's greatest spectacles, which had been overrun by tacky commercial enterprises. In 1872, Congress responded to this public concern by designating Yellowstone as a National Park, thus establishing an American tradition of parks and preserves that would serve as a model for the world.

The concept of National Parks was expanded in 1906 with the passage of the American Antiquities Act, which gave the President the authority to preserve, as "National Monuments," objects of cultural or historic significance on public lands. President Theodore Roosevelt broadened the original intent of the act in order to set aside spectacular natural areas. In the years that followed, the National Parks survived problems including lack of funding, short-sighted management, and conflicts between use and preservation—problems that continue today.

Yosemite National Park, located in central California, represents both the preciousness and vulnerability of our nation's parks. It is home to a host of natural wonders: powerful waterfalls, massive granite domes, and towering sequoias. Native Americans once prized the area for its unusual beauty, and the white explorers who supplanted them were awed as well. Today, its beauty has made Yosemite one of our country's most popular national parks, attracting millions of visitors every year. Unfortunately, the activities of these visitors, combined with the hotels, restaurants, and entertainment facilities designed to attract their business, are threatening to rob Yosemite of its wild beauty and sublime grandeur.

PHYSICAL BOUNDARIES: THE NATIONAL PARK SYSTEM AND YOSEMITE'S PLACE IN IT

Today's National Park System contains 50 National Parks, along with hundreds of other units such as National Monuments and Historic Sites, and covers almost 80 million acres. These units vary greatly in size, from the 8.3 million acre Wrangell-St. Elias National Park in Alaska to the one-third acre Ford's Theater National Historic Site in Washington, DC. The National Park System is managed by the National Park Service, which is part of the Department of the Interior. In creating the National Park Service, Congress decreed that the agency should manage the parks in order to conserve their scenery, wildlife, and natural and historic objects, and to provide for the enjoyment and use of the same by the public. Although the National Park System has expanded beyond the large western parks like Yellowstone and Yosemite which feature natural and scenic wonders, such parks are still considered the "crown jewels" of the National Park System.

Yosemite National Park covers 1189 square miles on the western edge of the Sierra Nevada mountain range (about the size of Rhode Island). The Sierras were formed approximately 25 million years ago, when the earth shifted, uplifting the granite ridges that had lain below its surface. Two or three million years ago, massive glaciers carved out valleys and lakes and sculpted granite peaks. One of the most striking legacies of the glaciers is the patches of glacially polished rock found in the upper Yosemite region; they are patches of granite so compacted and polished by the pressure of tons of ice that they shine like mirrors in the sun. Even today, small glaciers exist in the upper reaches of the Sierra Nevadas.

Because elevation ranges from 2000 to 13,000 feet in Yosemite, weather conditions can vary throughout the park on any given day. However, summers are generally warm and dry, and winters are usually mild, although 70-90 percent of the year's precipitation falls between November and March.

The park's most well-known area is the Yosemite Valley. Roughly seven miles long and one mile wide, the valley is contained between granite walls 900 feet high which appear to rise almost perpendicularly from the valley floor. Flowing through the valley is the Merced River, which forms some of the park's most spectacular waterfalls. Yosemite Falls, North America's highest waterfall, thunders down 2610 feet in three stages. Bridal Veil Falls, covered with the gauzy mist that gives it its name, often shimmers with rainbows. Ribbon Fall, also known as Virgin's Tears, alternates between a powerful cascade during the spring floods and a tear-like shower in the fall.

The Yosemite Valley also features some of the park's most impressive granite peaks, including El Capitan, the Cathedral Rocks, and the Three Brothers. Their massive size and elegant, sculpted lines give the valley the air of a cathedral.

BIOLOGICAL BOUNDARIES: PROTECTING PLANT AND ANIMAL SPECIES

National Parks, like other federal lands, are important not only as places for the public to enjoy the wonders of nature, but also as habitat for a variety of species. The parks owned by the federal government encompass an astounding diversity of living ecosystems—alpine meadows and prairies and coral reefs and mangrove swamps. These diverse ecosystems perform a myriad of ecological functions. Forests and grasslands, for example, hold soil in place, minimizing erosion and helping to keep streams and rivers clean and clear, while wetlands slow and absorb floodwaters, helping to protect adjacent land areas, and provide critical habitat for many species of fish and shellfish.

Because the public lands encompass a wide array of ecosystems, they also protect a diverse complement of living organisms. Preserving this diversity of species is important to continued progress in medicine, agriculture, and industry, all areas that make use of and learn from wild species. But many people are compelled to preserve biological diversity solely out of aesthetic considerations; for them, sharing the planet with the greatest possible diversity of living organisms is a key ingredient in securing and maintaining a high quality of life. Others argue that all species have a right to exist independent of their economic or aesthetic value to humans; preserving biological diversity is therefore an ethical or moral mandate.

Because of its large size and the variety of climates and elevations it encompasses, Yosemite National Park is home to an especially great number of species. Over 70 species of mammals, including elk, mule deer, and black bears, live in the park, which also contains at least 1200 species of flowering plants and 37 tree species. Eight of those plants are considered endangered under the Endangered Species Act, and 37 are designated rare by the state or the park.

Plant Species of Yosemite

Types of vegetation range from those commonly found in the California deserts and lowlands to the boreal or northern species also native to glacial regions in Canada and Alaska. The Foothill Belt, generally dry and warm, extends up to an average altitude of 3000 feet, and vegetation largely consists of shrubs and grass.

The Yellow Pine Belt extends between 3000 and 6200 feet and features a great variety of conifers, including the dominant yellow pine, the incense cedar, and a variety of willows along the streams. This zone is also home to three major groves of the famed sequoia, which range up to 250 feet in height, 30 feet in circumference, and 3500 years in age. The Mariposa grove boasts the Grizzly Giant sequoia, which at 2700 years old is the oldest known representative of its species. In addition, the Mariposa grove contains the famous Wawona Tunnel sequoia. A tunnel cut through its massive trunk in 1881 allowed traffic to pass through, and although the tree fell in 1969, its massive remains still lie in the grove.

Species of boreal origin, such as the silver pine and snow bush, thrive in the Upper Coniferous Belt, between 6200 feet and the timberline. Above the timberline (up to 13,090 feet), only plants that can withstand harsh winters and little warmth, such as the Arctic willow and alpine sorrel, can be found.

Animal Species of Yosemite

The many streams, rivers, and lakes in Yosemite are home to eleven species of fish, five of which occur naturally in the region. The other six are stocked for sport fishing. In 1972, as part of an effort to restore the park to a more natural condition, managers planned to stop stocking nonnative fish. However, the California Department of Fish and Game protested, and limited stocking continues.

Over 350 black bears live in the park, and problems sometimes arise because they tend to steal food and supplies from campers. The bears have an uncanny ability to open any sort of container and retrieve backpacks from branches high above the ground. The park is trying to resolve these problems by efforts that encourage bears to return to their original foraging habits. In addition to making refuse containers bear-proof and relocating problem bears to wilder areas, the Park Service has created visitor education programs and established rules that prohibit feeding the bears. Feeding them acclimates the bears to humans and can encourage the bears to raid campsites.

Although wolves once lived in Yosemite, there are none today, largely because of the efforts of the early Park Service to eliminate animals that might frighten visitors. Unfortunately, the wolf's absence has allowed populations of its natural prey, such as elk, to grow to such proportions that the animals starve during the winter, forcing the Park Service to kill off portions of these populations to prevent mass starvation. Another large predator, the grizzly bear, was eliminated in the 1890s.

The bighorn sheep, another of Yosemite's original inhabitants, lived primarily in the areas of higher elevation. However, disease, hunting, and destruction of parts of their habitat that extended outside the park eliminated the bighorn from Yosemite by 1914. The animal was restored in the park in March 1986, when the California Department of Fish and Game, the National Forest Service, and the National Park Service introduced a herd of 27 bighorns to the eastern boundary of Yosemite. The bighorns were brought by helicopter from one of two remaining herds of original Sierra Bighorn that live on Sand Mountain (on the east side of the Sierra Nevadas).

Yosemite is also home to several pairs of the endangered peregrine falcon. This species is threatened by the amount of the pesticide DDT which has accumulated in the food chain in and around Yosemite. Although DDT has been banned in the United States for over 20 years, residual DDT in California continues to pollute the Yosemite area. As the falcons ingest more and more DDT, the shells of their eggs become thinner and thinner. Because these thin-shelled eggs have little chance of hatching, the adults produce few offspring by themselves. To preserve the peregrine falcon population in Yosemite, the Park Service augments nests with thin-shelled eggs by replacing them with artificial eggs. Then, just before the time the eggs would have hatched, they replace the artificial eggs with two or three newly hatched falcons raised in captivity. The parent falcons then raise the babies as their own.

SOCIAL BOUNDARIES: CULTURAL SIGNIFICANCE AND CULTURAL THREATS

Like all National Parks, Yosemite is an area of great cultural significance. One way the Park Service preserves the cultural values of the park is through programs that educate visitors about the Indians who once lived in Yosemite, which they called Ahwahnee. The Indian Cultural Museum located in the park sponsors weekly demonstrations of Indian traditions and provides other cultural and historical background on the area. Yosemite also hosts an Annual Indian Days celebration, which includes activities demonstrating the traditions and cultural values of the area's Native American people. Although maintaining a sense of the culture that once flourished in the park's vicinity is an important function of Yosemite National Park, it is also an important symbol of our own culture's desire to preserve natural places. Ironically, it is that same culture that threatens the integrity of Yosemite and other National Parks.

Some threats our culture poses to National Parks are external, originating outside the parks' boundaries. One such threat is atmospheric pollution, which can be carried into the parks by wind patterns from urban areas, smelters, industrial plants, and coal-fired power plants. Another external threat is activities that occur on adjacent lands not protected by park designation. Activities such as mining and lumbering, and the roadbuilding that accompanies them, can create pollution and can seriously disrupt populations of wildlife that range across park borders. For example, destruction of portions of their habitat outside park boundaries contributed to the elimination of bighorn sheep from Yosemite.

Although external threats can affect any park, in Yosemite the worst dangers have usually originated within the park's borders. A recurring problem in Yosemite and all parks is a chronic lack of sufficient funding. The problem grew especially severe during the Reagan administration (1980-1988), when budget constraints resulted in reductions in permanent park personnel at many sites, and deteriorating conditions in nearly all the facilities.

Although Park Service Director Newton Drury was pressured into allowing tungsten mining within Yosemite during World War II, since the 1976 Mining in the Parks Act most parks are protected from mining. Unfortunately, any mineral claims filed before 1976 are valid as long as the claim is developed in accordance with federal regulations. Another major trend that threatens the biological integrity of our parks is the loss of species. Some of the parks are of an insufficient size to maintain viable populations of large mammals. In addition, predator species such as wolves and grizzly bears have been eliminated in many parks, including Yosemite. Unfortunately, without their natural enemies, game such as elk, deer, bison, and moose become too plentiful and overbrowse the vegetation of the area, which can no longer adequately support the increasing population. Consequently, many of the animals weaken and starve.

Of all the internal threats to our nation's parks, the most rapidly accelerating are overuse and overcrowding. The National Parks and Conservation Association, a citizen's group and self-appointed watchdog of the National Park Service, reported that throughout the mid to late 1980s, park visits increased about 4 percent annually, which averages out to roughly a few million more visits each year. In Yosemite, the number of visits has increased even more dramatically, partially because of the Park's proximity to the major cities of Los Angeles and San Francisco, and partially because of its famous beauty. Currently, nearly 2.5 million people visit the park each year.

In order to supply the needs and wants of the park's visitors over the years, a massive complex of commercial enterprises has sprung up. As of 1974, the Valley contained three restaurants, two cafeterias, one hotel dining room, four sandwich centers, one garage, two service stations, seven gift shops, two grocery stores, one delicatessen, one bank, one skating rink, three swimming pools, one miniature golf course, two tennis courts, 33 kennels, 114 horse and mule stalls, one barber and one beauty shop, and 13 liquor stores. Since that time, a shopping center has been constructed as well.

At first, these hotels, golf courses, and other facilities were thought of as valuable improvements to the park's appeal, but gradually the over-commercialized atmosphere detracted from visitors' experience of the park. In addition, the exponential growth in visitors which accompanied such "improvements" has led to other problems: heavily traveled trails have become eroded, vegetation has been trampled, and wildlife has been crowded into an ever shrinking space.

The visitors have also suffered. As their numbers have increased, individual visitors have found it more and more difficult to experience the essence of the park. During the popular summer months, overcrowding, excessive noise, and litter all detract from the peaceful atmosphere people hope to find in a National Park. Massive doses of campfire smoke and automobile exhaust pollute the air. In addition, crime has become a concern in recent years. Park rangers spend a good deal of their time dealing with speeders, drunk drivers, and petty thefts of purses, wallets, cameras, and radios; in 1985, rangers recorded 18,000 violations and made nearly 600 arrests.

LOOKING BACK: THE HISTORY OF YOSEMITE NATIONAL PARK

The first known sighting of the Yosemite area by non-Indians occurred in 1833, when the Joseph Reddeford Walker party passed by the region as they crossed from the east side of Sierras. However, the area was left undisturbed until gold was discovered in the Sierra Nevada foothills. The gold rush that began in 1849 led to armed conflicts between miners and the Native Americans whose homelands they threatened. As a result, the state of California sent an army battalion into the area to end what was known as the Mariposa Indian War. At that time, 22 villages of Ahwahneeches, a tribal branch of the southern Sierra Miwok, populated the area that now makes up the park. In 1851, the battalion chased a band of Ahwahneeches into the valley. The white soldiers were impressed with the beauty of the valley, and, within a few years, the army had "conquered" the area, killing or displacing all the Native Americans who had lived there.

Although the native population of Yosemite was eradicated in the mid-1800s, conflicts over the use of the area continued. Yosemite's history since that time has been one of a continual struggle between those who have wanted to develop the park for maximum public enjoyment and those who have tried to preserve the natural beauty and atmosphere of the environment.

How Public Use Grew from a Goal to a Problem

Once the conquering soldiers and the first few visitors spread the news of Yosemite's beauty and natural resources, the area was flooded by newcomers eager to make their fortunes. They gave little thought to their effect on the environment. Herds of cattle, sheep, and horses grazed meadows down to bare earth, and heavy logging also contributed to erosion. The roads, houses, stables, and gold-mining operations that began to encroach on Yosemite's wilderness brought the legacy of civilization: human activity, noise, and refuse.

Fortunately, conservationists headed by Frederick Law Olmstead and I.W. Raymond appealed to Congress to recognize Yosemite's value as a natural resource and to take steps to preserve it. In 1864, President Lincoln signed the bill that granted the Yosemite Valley and the nearby Mariposa Big Tree Grove of sequoias to the state of California, with the instruction that these areas be preserved specifically for public use and recreation.

One of Yosemite's earliest and most ardent admirers was the naturalist John Muir. He spent several years exploring the wonders of the area, which are described in his book *The Yosemite*. Muir exulted in the powerful natural forces of Yosemite, rushing out of his shelter to experience earthquakes, climbing tall trees to watch thunderstorms, and risking his life to achieve the best view of a waterfall. Partially in response to the herds of sheep (which he called "hooved locusts") that grazed the slopes of the Sierras down to the soil, Muir became part of the campaign to protect the area by make the area around Yosemite Valley a National Park. This goal was realized in 1890. In 1905, the state of California returned the valley itself to federal control so it could become part of the National Park as well.

Unfortunately, this triumph for conservation was shortly followed by a bitter defeat. The park's second major river, the Tuolumne, originally flowed through the Hetch Hetchy Valley, which was similar to Yosemite in its grandeur. When the city of San Francisco proposed damming up Hetch Hetchy to create a reservoir for the city, Muir and other conservationists fought to prevent the destruction of such a precious natural resource. But the city's wishes prevailed in 1913, and today the waterfalls and domes of Hetch Hetchy are buried under tons of water. According to friends, such a momentous defeat sapped Muir's will to live and, less than a year later, he died of pneumonia at the age of seventy-six.

The park was operated by the U.S. Army until 1917, when the newly created National Park Service assumed that duty. The early managers of the park did not share Muir's appreciation for the more violent side of nature. In fact, they tried to eliminate any factors that would prevent visitors from enjoying the park in peace and comfort. Naturally occurring fires were suppressed. Crews built roads and paths to areas of natural beauty. "Bad" animals that might discourage visitors, like wolves, were exterminated, allowing "good" animals that visitors enjoyed, like deer, to flourish. Stephen Mather, the first Director of the National Park Service, strongly encouraged people to visit Yosemite, and he ordered the construction of lodges and other "improvements" to attract them.

As time went on, the park became increasingly unlike the wild haven Muir had written about, and conservationists began to worry. The transformation of Yosemite and other parks especially concerned the Sierra Club, which Muir had founded in 1892 to promote the establishment and protection of National Parks. In the 1940s, they changed their Declaration of Purpose from "To explore, enjoy, and render accessible the mountains of the Sierra Nevada" to "To explore, enjoy, and preserve."

With the widespread use of the automobile, increased leisure time, and the dwindling amount of land remaining in a natural state, Yosemite and other parks became increasingly popular for vacations and week-end getaways. A few hours' drive from the metropolitan areas of Los Angeles and San Francisco, Yosemite experienced especially heavy influxes of visitors—too many, according to conservationists who worried about both the effect of so many hikers, campers, and automobiles on the natural environment and the quality of the visitors' experience. In fact, the contribution of automobiles in the park to pollution, noise and overcrowding had been predicted when they were first allowed in Yosemite in 1913. James Bryce, the British ambassador to the United States who visited Yosemite that year, warned, "If Adam had known what harm the serpent was going to work he would have tried to prevent him from finding lodgement in Eden; and if you were to realize what the result of the automobile will be in that wonderful, incomparable valley, you will keep it out."

The overuse problem came to a head on the weekend of July 4, 1970. Over 76,000 people visited the park that weekend—a record crowd. Several thousand young people gathered in a meadow, celebrating the holiday by playing loud music, smoking marijuana, and shedding their clothes. Park rangers asked them to leave, to no avail. Finally, a dozen rangers, armed with clubs and Mace, charged in, broke up the crowd, and arrested 186 people. Because of this incident, and other problems throughout the Park System, the NPS began a policy of training rangers in law enforcement. Although such measures were necessary, some people worried that as park rangers spent more and more time on crowd control and law enforcement, their roles as teachers, naturalists, and environmental managers would suffer.

How the Park Service Has Tried to Protect the Park

Soon after the July 4 incident, the Park Service began limiting visitors and eliminated auto traffic in the east end of the valley in an attempt to protect the park from environmental degradation and to preserve the wilderness experience for park visitors. Meanwhile, they worked on a master plan aimed at truly solving the park's problems. The proposed plan included the complete elimination of private automobiles in the park, the construction of a mass transit system, and the removal of all Park Service and concessioner buildings from the valley to the town of El Portal, at the western entrance to the valley.

This plan soon became the focus of a storm of controversy. In Yosemite, as in most other National Parks, all of the entertainment, restaurant, and lodging facilities had become consolidated under the management of a single concessioner. In Yosemite's case, the concessioner was the Yosemite Park and Curry Company. In 1973, this company was acquired by MCA, a large entertainment company that opposed any action that might limit its profits in any way. Conservationists charged that MCA had influenced the final master plan in favor of development rather than conservation, and persuaded the Assistant Secretary of the Interior to examine the plan.

In the meantime, MCA increased development in the already overcrowded valley, adding more facilities such as banks, motels, and boutiques. In addition, they attempted to produce a television adventure series about park rangers, but the show was so poorly produced that it made the park and the rangers look foolish. The final insult to the park occurred when the director of the series had some of the rocks in the park painted to improve their appearance on camera. This incident served as focus for conservationists already outraged at the commercialization of the park.

In the end, the Park Service's original master plan was deemed inadequate. Accordingly, they worked to revise it, incorporating the advice of over 60,000 concerned individual citizens. Eventually they developed a more comprehensive General Plan that was approved in 1980. The main goals of the plan were, in 10 or 15 years, to

• Designate 90 percent of the park as wilderness, forever free from development.

- Remove substandard NPS and concessioner staff housing and other facilities from Yosemite Valley to El Portal.
- Reduce concessioner-operated lodging facilities by 10 percent, and reduce overnight facilities in the valley by 17 percent.
- Reduce use of private vehicles in the valley, with a long range goal of eliminating them entirely.
- Identify and enforce carrying capacities (the goal for day use was 60 percent of 1980 levels, and for night use 85 percent).
- Improve and expand information, interpretation, and reservation services.

Identification and enforcement of carrying capacities, especially for overnight visitors, were an important steps in preserving the park. Unfortunately, the conservationist's goal—enhancing the wilderness experience for everyone by restricting the number of people who could enjoy it simultaneously—was misinterpreted by many people, who saw the restrictions as a blow to their individual freedom to enjoy Yosemite rather than a step necessary to preserve this national treasure. Some opponents of the plan, pointing to the removal of non-essential concessions from the park grounds (a move designed to restore the area's natural atmosphere and habitats) accused the conservationists and Park Service of limiting the enjoyment of Yosemite to the young and able-bodied.

In addition to changing its policies toward visitors, the Park Service altered the way it managed the vegetation and the wildlife of the park to produce a more natural ecosystem. One of the first steps was to end the artificial suppression of fires. To prevent natural fires from running rampant, the Park Service first had to eliminate an unnatural build-up of fuel, the legacy of decades of the no-burn policy. For instance, groves of sequoias, which would naturally be open, were clogged with an understory dominated by white pine. The Park Service began with a series of carefully controlled, purposefully set fires.

The Park Service also began a program to eliminate exotic plants brought into the park from other areas, using biological control agents whenever possible. They worked to revegetate sites stripped or altered by human activity and reduced the threat of further environmental degradation by regulating the grazing of horses, mules, and burros used for recreational trips or government work. Water quality monitoring was undertaken to safeguard against the pollution of lakes, streams, and rivers. The reintroduction of bighorn sheep and efforts to help breed peregrine falcons also add to the "re-naturalization" of the park.

Finally, the park has instituted a recycling center where visitors can return bottles and cans bought in the park for a return of five cents apiece. In addition, the center accepts bottles and cans from other sources, newspaper, and cardboard. In addition to reducing litter and the amount of garbage that must go into landfills, the project directs any profits to environmental projects within the park.

LOOKING AHEAD: THE FUTURE OF YOSEMITE NATIONAL PARK

Although the Park Service has temporarily safeguarded Yosemite from the effects of crowds of people—especially noise, pollution, and development—the park's future remains in danger because our nation cannot resolve the controversy over use versus preservation.

When the National Park Service was established in 1917, Congress defined its primary purposes as providing for the enjoyment of the public and conserving the scenery, natural and historical objects, and wildlife within the parks in ways that would preserve them unimpaired for the enjoyment of future generations. In other words, there are limits on how we the public may use "our" National Parks—we may not build, develop, or alter them in ways that will prevent others from experiencing the same peaceful environment and breathtaking sights which we enjoy today. This philosophy underlies the limits on visitors and the removal of concessions from the valley, and should continue to inform management of the park.

In 1992, the Park Service plans to begin the process of moving a number of its maintenance and warehouse operations to El Portal. Once the move is accomplished, they plan to restore five acres in the valley to their natural condition, which is generally a black oak woodland. The Park Service estimates they will complete this project by 1999, at a cost of $4 million.

Although this plan is a valuable sign of progress, no steps have been taken to move nonpark facilities from Yosemite Valley to El Portal. Concessioners protest moving because of the money they have invested in their current location. Although concessioners should certainly be allowed to operate outside park limits, their profits should not be the major concern. In fact, the Park Service legally owns any facilities constructed on park property and should be able to force the removal of these operations. Only a few facilities essential for lodging and feeding visitors should remain inside the park.

Finally, the Park Service has yet to take action to significantly reduce the amount of automobile traffic in the park. Many critics believe that this is the most effective step they could take to reduce noise, pollution, overcrowding, and other undesirable effects that harm the Yosemite environment as well as visitors' experience of the park.

Another promising initiative that could help protect Yosemite is the National Park System Protection and Resources Management Act (commonly called the Park Protection Act), which was introduced to Congress in the early 1980s. The major thrust of the bill was twofold: to establish ways in which park administrators can track the condition of the natural and cultural resources within park boundaries, and to give the Park Service a stronger voice in determining how other federal agencies use or affect park lands.

The House of Representatives, in two different Congresses, has passed the Park Protection Act, but the Senate has successfully blocked its passage. Many park advocates warn that if steps are not soon taken to provide adequate protection, the parks face the probability of continued degradation and irrevocable change, change that will diminish or destroy the very qualities that make these areas worthy of National Park status. However, given the increased environmental awareness of the general public, and their enthusiastic support of the National Parks, the bill will once again be debated before Congress, perhaps with a different and more favorable outcome.

In his 1912 argument against damming the Hetch Hetchy Valley, John Muir wrote that "no holier temple has ever been consecrated by the heart of man." Preservation, which includes limits on use, must be the top priority in order to preserve for future generations the natural temples of Yosemite and other National Parks .

SUGGESTED READING

Byrnes, Patricia. "Restoration is the Goal: A Trust for Yosemite." *Wilderness*. Winter 1990. 4-6.

"Control of the Wild." *National Parks*. September/October 1990. 20.

Darling, F. Frazer, and Noel D. Eichhorn. *Man and Nature in the National Parks*. Washington, DC: The Conservation Foundation, 1971.

Foresta, Ronald A. *America's National Parks and Their Keepers*. Washington, DC: Resources for the Future, 1984.

Freemuth, John C. *Islands Under Siege: National Parks and the Politics of External Threats*. Lawrence, KS: University Press of Kansas, 1991.

Hartzog, G.B., Jr. *Battling for the National Parks*. Mt. Kisco, NY: Moyer Bell Ltd., 1988.

Huber, N. King. *The Geologic Story of Yosemite National Park*. Washington, DC: U.S. Geological Survey, 1987.

Ise, John. *Our National Park Policy, A Critical History*. Baltimore: Johns Hopkins University Press, 1961.

Jones, Holway R. *John Muir and the Sierra Club: The Battle for Yosemite*. San Francisco: Sierra Club, 1965.

Koshland, Daniel E. "The Saving of Yosemite." *Science*. February 15 1991. 721.

McNeely, Jeffrey A. "The Future of National Parks." *Environment*. January/February 1990. 16+.

Mitchell, John. "Uncluttering Yosemite." *Audubon*. November 1990. 72-80+.

Muir, John. *Our National Parks*. Madison, WI: University of Wisconsin Press, 1981.

———. *The Yosemite: The Original John Muir Text*. San Francisco: Sierra Club Books, 1989.

Rauber, Paul. "Yosemite: Paradise Regained?" *Sierra*. March/April 1991. 24+

Reinhardt, Richard. "Careless Love." *Wilderness*. Summer 1989. 17-27.

Runte, Alfred. *The National Parks: An American Experience*. Lincoln, NE: University of Nebraska Press, 1979.

———. *Yosemite: The Embattled Wilderness*. Lincoln, NE: University of Nebraska Press, 1990.

Russel, Carl Parcher. *One Hundred Years in Yosemite: The Story of a Great Park and Its Friends*. Yosemite National Park: Yosemite Natural History Association, 1957.

Sax, Joseph L. *Mountains without Handrails: Reflections on the National Parks.* New York: Alfred A. Knopf, 1951.

Scharff, Robert, ed. *Yosemite National Park.* New York: D. McKay Co., 1967.

Tilden, Freeman. *The National Parks.* New York: Alfred A. Knopf, 1983.

Zaslowsky, Dyan, and the Wilderness Society. *These American Lands: Parks, Wilderness, and the Public Lands.* New York: Henry Holt and Company, 1986.

CHAPTER *19*

WILDERNESS: THE BOUNDARY WATERS CANOE AREA WILDERNESS

Wilderness—for some, the word is intimidating. The idea of vast expanses of land without roads, electricity, or telephones, where humans are as subject to the forces of nature as any other animal, may frighten those who have never experienced it. Yet the preservation of wilderness is essential to all of us, not just to people who feel comfortable backpacking through pathless forests or canoeing for days without the sight of a building or a bridge. Wilderness areas contain healthy, unaltered ecosystems that help to purify air and water, provide homes for endangered species, and perform numerous other important functions that we often take for granted. In addition, for citizens of the United States in particular, wilderness is an essential element of our cultural heritage; the fact that our country contains expanses of untouched natural areas has shaped the way we think of ourselves and our relationship with nature. Finally, many people believe our society has a psychological need and a moral duty to preserve at least some parts of the environment completely undisturbed by civilization.

The United States realized the importance of preserving natural areas as early as 1832, when Congress declared the region of Hot Springs, Arkansas, the first natural federal preserve. Today natural areas are protected through our systems of National Parks, National Forests, and Bureau of Land Management areas. However, for an area to be considered true wilderness, it must be minimally disturbed by humans, and many of these protected areas have been altered by the construction of roads, logging, grazing, or other human activities. Concern for the untouched, true wilderness areas of our country has been strong since at least 1935, when the Wilderness Society was formed in response to threats to such areas, especially the reluctance of land-managing agencies like the Forest Service and the Park Service to protect wilderness from powerful economic interests.

The first officially designated wilderness areas were located exclusively within the National Forests and Indian Reservations, and there was no standard procedure for managing them. It was not until 1957 that Hubert H. Humphrey, the Democratic Senator from Minnesota, introduced a bill that would spell out the legal definition of wilderness and establish basic standards for designating and managing wilderness areas. Despite opposition from commercial interests (primarily logging, mining, and ranching) with significant political clout and, ironically, the National Park Service and the Forest Service, the bill was passed in 1964 as the Wilderness Act. According to the act, wilderness is "an area where the earth and community of life are untrammeled by man, where man himself is a visitor who does not remain."

Despite the unprecedented protection granted to wilderness areas by the Wilderness Act, they are still vulnerable to threats from outside their borders, such as pollution and encroaching development. Confusion over which agency controls wilderness areas can also lead to inadequate protection. In addition, wilderness areas are sometimes harmed by those who love them the most: their visitors.

One popular wilderness area that exemplifies many of the benefits and controversies of wilderness preservation is the Boundary Waters Canoe Area Wilderness, which stretches for 120 miles along the Minnesota-Canada border. The BWCA, as it is called, provides a unique wilderness experience for the canoeists and campers who explore its vast network of lakes, rivers, and forests. Although the United States recognized the value of preserving this area in its natural state as early as 1902, its history is filled with conflicts over its proper use and confusion over who should manage it. Throughout the twentieth century, conservationists have battled over the fate of the BWCA with those anxious to develop the area for mining, logging, or recreation. Even today, although the BWCA is federally protected as wilderness, problems such as overuse, acid rain, and mining outside the boundaries threaten to degrade the wilderness environment and compromise its natural beauty.

PHYSICAL BOUNDARIES: THE DISTRIBUTION OF U.S. WILDERNESS AREAS AND THE ROLE OF THE BWCA

When Congress passed the Wilderness Act of 1964, it designated nine million acres of federal land as protected wilderness. These were the original constituents of the National Wilderness Preservation System. By the late 1980s, the Wilderness Preservation System had grown nearly tenfold, to over 89 million acres in 445 areas. Most of the wilderness areas lie within national park boundaries, with national forests, wildlife refuges, and Bureau of Land Management lands hosting the remaining areas.

Most protected wilderness areas are high alpine or tundra ecosystems, and most are in the western United States or Alaska. Fifty-six million acres, well over half of all designated wilderness acreage, lie in Alaska alone. But two thirds of all recreational uses in the National Wilderness Preservation System occur on just 10 percent of wilderness areas in the lower 48 states, especially in California, North Carolina, and Minnesota. In the contiguous United States, where the demand for it is the greatest, protected wilderness areas comprise less than 2 percent of the total land area. By conservative estimates, another 5 percent qualifies for wilderness designation. Even if this land were added to the National Wilderness Preservation System—an unlikely event—less than 8 percent of the contiguous United States would be designated wilderness.

The largest wilderness area east of the Rocky Mountains, the BWCA Wilderness spreads over 1,030,000 acres along the northern reaches of the Superior National Forest. Stretching from the northern shore of Lake Superior near Grand Portage State Forest to Voyageur National Park in the west, it is the nation's largest canoe wilderness.

The BWCA's most outstanding feature is its intricate tangle of more than 1000 lakes linked by hundreds of miles of streams and short portages. These waterways once served fur traders and the Sioux and Chippewa people who navigated the canoe country hundreds of years ago. Bordering the BWCA to the north is Canada's Quetico National Forest. Together, these two protected areas create a canoeist's paradise: over 3000 square miles of undisturbed woodland, islands, rivers, and lakes.

The BWCA owes much of its unusual nature to its geological formation. The wilderness occupies the southern edge of the Canadian Shield, a Precambrian formation extending from central and eastern Canada to the upper midwest and New York. This formation was created over three billion years ago, when volcanic lava flows solidified, and then were uplifted and eroded. With the onset of the Ice Age, glaciers carved out the vast network of lakes and rivers that we see today.

BIOLOGICAL BOUNDARIES: THE ROLE OF WILDERNESS AND THE BWCA IN PRESERVING SPECIES DIVERSITY

Although the Endangered Species Act helps to protect individual species from extinction, scientists agree that we can only save a small percentage of endangered plants and animals by focusing on individual species. Clearly, our best chance to prevent more species from becoming endangered and to save the greatest possible number of species is by preserving the variety of ecosystems that are their habitats. Although wilderness designation would obviously preserve an ecosystem in an undisturbed state, our National Wilderness Preservation System falls woefully short of protecting the full range of ecosystems in our country. Scientists have identified 233 distinct ecosystem types in the United States, yet only 81 are represented in the NWPS. Fifty are no longer on federal lands, and so cannot be officially protected. The remainder—some 100 ecosystem types—do occur on federal lands but are not now protected. How much longer they can retain their integrity without official protection is unknown.

The preservation of ecosystems in their natural state not only provides species with undisturbed habitat, but also helps protect them in other ways. Scientists can study undisturbed areas to determine how human actions have altered similar, but unprotected, ecosystems. Using the wilderness ecosystem as a reference, they can gauge the effect of cultural actions and the extent of resultant changes in the modified ecosystem. Further, wilderness areas may provide scientists with the clues and understanding required to restore the ecological health and integrity of altered ecosystems.

The BWCA plays an important role in preserving species and their habitat. Because of its vast size and close proximity to relatively untrammeled areas in Canada, the BWCA is home to species of wildlife like the timber wolf, largely exterminated in other areas of the United States. The BWCA also hosts a multitude of other mammals, including moose, deer, beavers, snowshoe hares, porcupines, eastern timber wolds, pine marten fishers, and lynx. Birdwatchers can find hermit thrushes, white throats, partridges, loons, and even bald eagles, while those who fish can enjoy such sportfish as bass and pike.

Although extensive logging occurred in the BWCA area between 1895 and the time it became protected by wilderness designation, roughly 540,000 acres of virgin forest remain today, by far the largest area of forest untouched by logging in the eastern United States. These forests include stands of tall red and white pine, redolent cedars, and silvery birches.

SOCIAL BOUNDARIES: THREATS TO WILDERNESS FROM ECONOMIC ACTIVITIES AND OVERUSE

Although wilderness designation protects natural areas from exploitation to an even greater degree than designation as a National Forest or park, even wilderness areas are still threatened by some economic activities and by overuse.

In general, wilderness areas today are open only for recreational activities such as hiking, sport fishing, camping, and nonmotorized boating. In some areas, sport hunting and horseback riding are also allowed. Motorized vehicles, boats, and equipment are prohibited unless they are required in an emergency, such as fire control or a rescue attempt. Although roads, human-made structures, timber harvesting, livestock grazing, and mining are also prohibited, preservationists were forced to compromise on several points in order to secure congressional support for the wilderness bill. Mining and grazing are permissible if they were initiated before the area was designated as wilderness. In addition, the exploration and development of mining, prospecting, and oil and gas drilling were allowed to continue at the discretion of the Secretary of the Interior until December 31, 1983, and any claims filed by that date can still be developed at any time in the future.

Although, like all wilderness areas, the BWCA is vulnerable to economic exploitation allowed by these exceptions in the Wilderness Act, perhaps the greatest threat to its integrity as a wilderness is its popularity. Today, the BWCA is the most heavily visited unit of the entire wilderness system, drawing people from throughout the country. In 1987, the area received 180,000 visitors—a 10 percent increase from the previous year. Although the BWCA places limits on visitors during the summer, it still receives ten times as many visitors as does Quetico National Forest, which imposes much stricter quotas. In addition, the BWCA experiences problems due to uneven concentrations of use. Overnight visitors concentrate in the 2200 designated campsites. Over 50 percent of the groups which enter the BWCA do so by one of seven entry points, while the other 80 entrances go relatively unused. The most popular portages are subject to erosion and crowding, thus damaging the environment as well as detracting from the wilderness experience.

Despite ongoing threats to its unspoiled nature, the BWCA continues to offer diverse benefits to the society which both protects and endangers it. In addition to providing a haven for canoeists and campers, it serves as a valuable educational and scientific resource. The wilderness area's rich ecosystem has been the focal point of research in wildlife behavior, acid deposition, forest ecology, nutrient cycles, lake systems, and vegetation history.

LOOKING BACK: THE HISTORY OF THE BOUNDARY WATERS CANOE AREA

Since the early 1900s, the best use of the area now known as the BWCA Wilderness has been the subject of controversy: should its resources be open to private industry and individual ownership, or should the unique area be preserved in its wild state for public enjoyment and as a standard of true wilderness?

Managing the BWCA Before the Wilderness Act

Shortly after the turn of the century, a growing public concern over watershed protection and similar conservation issues surfaced. This concern lay behind the 1902 decision of the U.S. Land Office to set aside 500,000 acres in the area now known as the BWCA. This land was intended as a forest reservation, protected from settlement and development. The Land Office withdrew an additional 659,000 acres in the same area by 1908.

In 1909, President Teddy Roosevelt designated 1.2 million acres of the previously withdrawn area as the Superior National Forest. Quetico Forest, established the same year by the Canadian government, achieved park classification within four years. Superior would have to wait much longer to receive the similar protection afforded by wilderness designation.

Management of the area was confused and complicated from the beginning. Arthur Carhart, a landscape architect who visited Superior in 1919 and again in 1921, recommended that it should be kept as close to wilderness as possible. In 1926, Agriculture Secretary William Jardine halted plans by local governments to build roads to many of the lakes. He established a policy to protect the wilderness quality of Superior by prohibiting unnecessary road construction and preserving a 1000 square mile area with no roads whatsoever.

But other interests, especially hydropower developers, had their eyes on Superior. Proposed dams would have raised some lake levels 80 feet, drowning scenic rapids and waterfalls while requiring the construction of multiple roads. Conservationists pressured Congress to protect Superior from these ravages by law, and in 1930, the Shipstead-Newton-Nolan Act was passed. This act—the first in U.S. history specifically designed to protect wilderness—prohibited logging within 400 feet of lakeshores, forbade any alteration of natural water levels (such as those caused by hydropower dams), and withdrew public lands in the area from homesteading. Hydropower interests unsuccessfully tried to have the act repealed.

Throughout the 1930s, the federal government bought tracts of land and repossessed tax-delinquent property, adding to the protected acreage in Superior National Forest. The Civilian Conservation Corps built portage trails on portions of the newly acquired land as part of President Roosevelt's plan to create jobs during the Depression. Unfortunately, the government frequently lacked funds to purchase pieces of land especially suited or crucial to the wilderness area.

After World War II, resorts sprang up on privately owned land in Superior. These resorts were frequently accessed by floatplanes, which conservationists felt insulted and disturbed the wilderness environment. Although the wilderness area became known as the Superior Roadless Primitive Area in 1938, logging operations soon grew up around its perimeter. In addition, some logging companies pressured the Forest Service into allowing them to make inroads into the roadless area itself.

To help protect Superior from commercialization, concerned individuals sometimes assisted government efforts to acquire privately owned land. For example, beginning in 1937, lawyer Frank Hubachek contributed hundreds of acres. In 1943, the Izaak Walton League joined such efforts. These groups and individuals gave or sold (at a loss) many key holdings to the Forest Service. In 1948 Congress helped the effort by passing the Thye-Blatnik Act, which gave the Forest Service authority to purchase private inholdings scattered throughout the wilderness area. Just as importantly, Congress provided $9 million to fund the purchase of private lands.

In 1949, President Truman also joined the preservation effort by establishing an unprecedented 4000-foot airspace reservation over the wilderness area, effectively preventing floatplanes from landing on wilderness lakes. Although fishers and hunters protested this order as unconstitutional, a District Court of Appeals backed Truman's policy.

The protection of the BWCA was fought by a consortium of powerful interests who wanted to develop the area: logging and mining companies, resort owners, and people who wanted to enjoy "nature" in mechanized comfort. These groups accused Hubachek and other conservationists of wanting to create their own private wilderness or to increase the value of their own land holdings. Despite the controversy, in 1954 Congress extended the Thye-Blatnick Act to include almost all of the present BWCA, and provided $2.5 million for land purchases. In 1958, the Superior Roadless Area officially became known as the Boundary Waters Canoe Area.

How the Wilderness Act Affected BWCA

Despite the apparent support of Congress, conservationists soon realized that the wilderness quality of Superior National Forest remained in danger. The Forest Service seemed unable to decide if logging or preservation was the higher priority, and confusion over what was and was not permitted in the BWCA abounded.

The BWCA was one of the original areas included in the National Wilderness Preservation System, despite the fact that it was not "untrammeled" by human activity, as stipulated by the 1964 Wilderness Preservation Act. The area had endured decades of use and abuse by trappers, loggers, commercial fishers, and resort owners. However, the efforts of those committed to preserving the BWCA's wilderness qualities, combined with the area's unique beauty and vast stretches of still untouched wild lands, earned the area its wilderness designation.

With the passage of the 1964 Wilderness Act, the BWCA became a key unit in the new National Wilderness Preservation System. Unfortunately, the act did not end confusion over use of the BWCA because it contained sections allowing logging and motorized vehicles to continue in some areas of the Boundary Waters.

In a compromise effort to appease conservationists and developers, in 1965 the Secretary of Agriculture proposed a management plan that divided the BWCA into two zones. The Interior Zone contained over 618,000 acres and 90 percent of the water area (and thus the bulk of the recreational use). Logging was strictly prohibited in this zone, but allowed in the Portal Zone, the remaining 412,000 acres of the BWCA.

In 1972 Minnesota Public Interest Research and the Sierra Club sued the Forest Service and several logging companies in an attempt to protect virgin forests in the Portal Zone. The U.S. District Court that heard the case sided with the conservationists, but on August 30, 1976, the federal appeals court reversed the ruling to permit logging. The appeals court added that virgin timber cutting in the BWCA could only be prevented by substantive changes in policy made by Congress or the Forest Service.

Additional lawsuits sprang up over similar management issues such as mining and motorized vehicles. Too frequently, those who wanted to preserve the wilderness character came from cities like St. Paul, Minneapolis, Chicago, and St. Louis, while those who favored development and motorized recreation lived near the BWCA and often earned their livings as resort owners or fishing trip guides; the conflict became regional and socioeconomic as well as use-oriented.

In 1978, Congress finally settled these management conflicts by adopting the BWCA Wilderness Act, which ended logging and road-building in any area, severely restricted mining, and provided funding to buy property from resort and lodge owners. The act also cut motorboat use from 60 percent of the water area to 33 percent, with the eventual goal of phasing it out altogether. To protect and enhance the natural qualities of the area, the act called for restoring natural conditions to areas damaged or altered by human activity. Congress tried to compensate businesses and resorts for the removal of operations from the BWCA by providing more facilities outside its borders. The government also offered to assist the local merchants by purchasing resorts that became unprofitable as a result of the act.

How BWCA's Visitors Almost "Loved the Wilderness to Death"

Although the BWCA Wilderness Act gave the area substantial protection from outside threats, it did not address an equally serious problem: the increased demand for authentic wilderness experiences by the public. By the early 1970s, the problem of overuse had grown so severe that managers worried that it would not only detract from the wilderness experience, but also permanently degrade areas of the BWCA ecosystem.

In 1976, the Forest Service initiated a visitor distribution program to control overuse. They used travel simulator models to place daily limits on the number of visitors who could enter the wilderness. The computer models, which simulated lake-to-lake movements by campers, recommended optimum entry rates based on predicted occupancy levels of lakeside campsites. At first this user control program encountered opposition from commercial interests in the area, including canoe outfitters, resorts, and camps. Eventually, however, these groups were persuaded to support a program of gradual adjustment to recommended entry levels. The entry program went in to effect in 1979.

In order to enforce the entry quotas, all visitors between May 1 and September 30, the most popular season to visit the BWCA, must have entry permits. These permits are free, but can also be reserved ahead of time for a $5 charge. Another requirement that controls overuse is a party limit of ten during the May-September season; larger parties cannot receive permits.

Current Environmental Threats to the BWCA

Today, the U.S. government is firmly committed to maintaining the BWCA as a true wilderness area, and a network of laws and management programs help work toward this goal. This network is necessary to protect the BWCA from numerous external and internal forces.

External Forces. Although wilderness areas are relatively protected from mining, operations that occur outside the area's borders can still affect the wilderness environment. Pollution of air and water knows no boundaries, and neither do animals; disruption of animal habitats outside the wilderness frequently disturbs populations that roam in and out of the protected area.

Copper-nickel-sulfide ores and precious metals are found in and around the BWCA, and potential mines outside its borders pose a real threat to the integrity of the BWCA wilderness.

Only by expanding the wilderness area or establishing a "buffer zone" where damaging activities like mining are prohibited can we adequately protect the BWCA.

Although President Truman's airspace reservation saved BWCA from the intrusion of floatplanes, a new aeronautical threat emerged in later years. A military airspace called the Snoopy Military Operations Area has been established over the BWCA, just above the protected airspace. Military jet flights, which increased in frequency by ten times between 1983 and 1986, produce sonic booms and exhaust vapors that could disturb animal populations, inhibit plant growth, and generally detract from the wilderness experience. The military has conducted virtually no evaluations of how these flights affect the BWCA, although conservationists claim that studies of similar problems reveal a significant potential for environmental damage. In 1988, the Friends of the Boundary Waters Wilderness (FBWW) and other conservation groups initiated legislation to prevent use of the airspace until adequate environmental reviews have been conducted, but the outcome of this action is still unknown.

Another growing problem is the effect of acid rain on the lakes and forests of the BWCA. Because aquatic ecosystems are believed to be most sensitive to acid deposition, BWCA's thousand lakes and related waterways make it especially vulnerable, and the wilderness' vast tracts of forest are at risk as well. In fact, a Minnesota study rated BWCA as the area in the state most sensitive to acid deposition. In an attempt to combat the problem, Minnesota passed the Acid Deposition Control Act in 1982, which directed the state to identify areas especially sensitive to acid deposition, set standards for those areas, and devise controls to maintain those standards. The 1990 reauthorization of the Clean Air Act places stricter standards on the emission of substances that contribute to acid precipitation nationwide, but only time will tell how past and ongoing acid deposition has affected the BWCA's lakes, streams, and forests.

Other external threats include a proposed National Guard Training Facility near the border of the BWCA. This facility would include areas for tank maneuvers, artillery ranges, and other activities likely to impinge upon and degrade the wilderness environment. Power plants, logging operations, and other commercial operations outside the BWCA can harm the wilderness as well. Fortunately, groups like the Friends of the Boundary Waters Wilderness sponsor citizen wilderness surveillance programs, in which concerned individuals uncover, report, and combat activities or developments that might harm the BWCA.

Internal Forces. Although the quota system has helped protect the BWCA from overuse, even a modest number of visitors can destroy vital components of the area's ecosystem if they do not understand and respect the wilderness environment. The Forest Service has found that well-publicized regulations, such as the prohibition of glass or metal containers, helps to solve this problem. Regulations are even more effective when explained by a uniformed aide; a 1981 program in which an aide passed out and explained a brochure on tree damage at a campground reduced tree damage by 81 percent. The Forest Service must increase programs that prevent visitors from damaging the environment.

The Forest Service must also repair what damage does occur. Besides replanting trees that were lost in logging operations, they are considering a program to reintroduce woodland caribou, a species that was driven out of the BWCA by human activity. They also perform smaller but equally necessary tasks such as rebuilding canoe entry points to prevent erosion.

LOOKING AHEAD: THE FUTURE OF THE BOUNDARY WATERS CANOE AREA WILDERNESS

In the past few decades, the National Forest Service, which manages the BWCA, has accomplished a transformation: many areas previously scarred by humans—such as logged areas that have been replanted with indigenous trees—have now returned to almost pristine conditions. The federal government also bought up private holdings in the wilderness, eliminated resorts and private cabins, and banned mining, logging, and other commercial activities as well as the use of snowmobiles, chainsaws, and machines in general. Even airplanes and outboard motors are restricted.

Conservation groups and other concerned citizens are vital to the health of the BWCA wilderness. The Friends of the Boundary Waters Wilderness played a major part in discontinuing the use of eight small dams in the area, establishing pesticide regulations, and a program to convince the owners of mineral rights within the BWCA to give up those rights. Another group, the Boundary Waters Wilderness Foundation, has been instrumental in initiating acid rain studies and the proposal to reintroduce woodland caribou.

In order to encourage this type of public involvement, the Forest Service conducts a variety of public information programs. For instance, one program uses a slide/tape show to instruct visitors on proper wilderness behavior when they pick up their entry permits. The Forest Service, as managers of the BWCA, could expand educational efforts by initiating activities like schools tours and weekend workshops. They could also encourage universities in the area to offer courses in wilderness management and community education classes on wilderness issues.

Although public support is crucial to the maintenance of BWCA, the area must also have a management structure that is efficient and effective if it is to endure as a lasting, protected wilderness. Currently, different aspects of the BWCA fall under the authority of multiple groups, including the National Forest Service, the Fish and Wildlife Service, the Minnesota legislature, and Congress itself. Currently, any one group seldom knows what any of the others is doing, leading to conflicts and costly delays on important actions. Establishing an independent review board to coordinate and evaluate the actions of all these agencies would help make decisions affecting the BWCA more efficient. Another option would be to bring the involved agencies together into a compact organization specifically designed for managing the BWCA. One way or another, management must be coordinated if inaction and confusion among well-meaning parties are not to be the downfall of this hard-won wilderness area.

SUGGESTED READING

"A Wilderness Vacation in a Canoe" *Changing Times*. June 1982. 15-16.

Allin, Craig W. *The Politics of Wilderness Preservation*. Westport, CT: Greenwood Press, 1982.

Blacklock, Craig. *Border Country: The Quetico-Superior Wilderness*. Minocqua, WI: North Word Press, 1988.

"Boundary Waters." *Wilderness*. Winter 1984. 26.

"Boundary Waters Canoe Area: We Like it Wild." *National Parks and Conservation Magazine.* October 1977. 20-27.

Brooks, Paul. "A Roadless Area Revisited." *Audubon.* March 1975. 28-37.

"Confusion at Boundary Waters." *National Parks and Conservation Magazine.* January 1977. 12-16.

Friends of the Boundary Waters Wilderness. *Outline History of the BWCA.* Minneapolis, MN: 1982.

Frome, Michael. *Battle for Wilderness.* New York: Praeger Publishers, 1974.

——, ed. *Issues in Wilderness Management.* Boulder, CO: Westview Press, 1985.

Harvey, Michael. "Wilderness Legislation Released by Compromise." *Environment.* December 1984. 34-37.

Hendee, John C., George H. Stankey, and Robert C. Lucas. *Wilderness Management.* U.S. Department of Agriculture. Washington, DC: 1978.

Hulbert, James. "BWCA Visitor Distribution System." *Journal of Forestry.* June 1977. 338-340.

Irland, Lloyd. *Wilderness Economics and Policy.* Lexington, MA; Lexington Books, 1979.

McCloskey, Maxine, and James Gilligan, eds. *Wilderness and the Quality of Life.* San Francisco: Sierra Club, 1969.

Mitchell, Lee Clark. *Witnesses to a Vanishing America.* Princeton, NJ: Princeton University Press, 1981.

Proescholdt, Kevin. "BWCA: The Embattled Wilderness." *American Forests.* July/August 1989. 29-78.

Searle, R. Newell. *Saving Quetico-Superior: A Land Set Apart.* Minnesota Historical Society, 1977.

Turner, Frederick. *Beyond Geography: The Western Spirit Against the Wilderness.* New York: Viking, 1981.

Wheelwright, Jane Hollister, and Lynda Wheelwright Schmidt. *The Long Shore: A Psychological Experience of Wilderness.* San Francisco: Sierra Club Books, 1991.

CHAPTER *20*

BIOLOGICAL RESOURCES: THE ST. LOUIS ZOOLOGICAL PARK

When most of us think of zoos, we think of places we go to enjoy seeing exotic animals like elephants and giraffes. Although zoos do strive to be entertaining, what we may not realize is that visitors also learn a great deal at today's zoological parks. In addition to observing what the animals look like and how they behave, they may learn about the animals' native habitats by reading posted signs or by examining the design of their enclosures. Perhaps the most crucial role zoos play today is to help preserve endangered species. As more and more of the Earth's animals become threatened by extinction, the zoo has become an increasingly important sanctuary for such species. Although zoos alone cannot halt the widespread extinction of animal species—only saving their native habitats from agriculture, development, logging, poaching, pollution, and other threats can do that—zoos can help, not only by educating the public about the animals' plight, but also by breeding endangered species in the hope that someday they can safely be returned to the wild.

Although not as large or well known as some other zoos, the St. Louis Zoological Park in St. Louis, Missouri, is an excellent example of how zoos all over the world are making important contributions to the preservation of endangered species. Through a consistent effort to stay on the forefront of progress, the St. Louis Zoo has accomplished the conservation-oriented activities of breeding, conducting research and educating visitors while retaining the purpose of entertaining the public.

PHYSICAL BOUNDARIES: THE LIMITS OF ZOOS AND THE ST. LOUIS ZOO

To many zoo visitors, the variety of mammals, insects, reptiles, and other animals is amazing. Compared to the number of animals species that exist, however, the number that can be maintained in zoos is miniscule. The St. Louis Zoo is no exception, containing roughly 2300 animals overall.

The Zoo occupies 83 acres in Forest Park, a recreational and cultural center near downtown St. Louis. As visitors approach the zoo, they can see the impressive walk-through flight cage originally constructed for the 1904 World's Fair, an event that helped encourage public support for the creation, in 1914, of the zoo itself. Visitors can enter the zoo free of charge and, once inside, enjoy a variety of displays and activities: 15 major animal exhibits, a children's zoo, an innovative new education center, three facilities for chimp, sea lion and elephant shows, gift shops and refreshment stands, and a train that makes four stops throughout the park.

BIOLOGICAL BOUNDARIES: EXTINCTION, PRESERVATION, AND THE ROLE OF ZOOS

Although zoos alone cannot halt the extinction of animal species, they do play an important role by providing a refuge for endangered species until their natural habitats can be safeguarded.

The Need to Protect Species in Danger of Extinction

Extinction is an issue that has received a great deal of publicity and is the subject of some controversy. Some skeptics claim extinction is nothing to be concerned about because it is a natural process that has occurred throughout Earth's history. Although this second point is correct—in fact, many experts believe that only 1-2 percent of all the species that have ever lived on earth are in existence today—the first is not, because human activities have greatly accelerated the rate of extinction, perhaps to as much as 400 times greater than normal. The greatest threat to most species is loss of habitat due to pollution, development, agriculture, logging, and other activities that drain wetlands, burn or cut forests, and plow or pave over grasslands. Illegal poaching is another major factor that threatens animal species from snakes to elephants.

Most people instinctively hate the idea that any entire species might vanish from the face of the earth. We naturally value areas that are biologically diverse, such as the national parks and forests, where so many of us go on vacation. We delight in spotting unusual species of birds or mammals, and photographing scenes that are aesthetically pleasing. However, the desire to protect the earth's variety of species, or biodiversity, from human activities that threaten their survival is more than an aesthetic luxury. The disappearance of an individual species—plant, animal, or bacteria—almost always has far reaching effects. Each species plays a distinct role in its ecosystem, and when it is gone, the organisms that depended on it may be threatened as well. For example, there are over 600 types of fig trees in the tropical rain forest, and figs are an important food for many of the animal species that live there. Each type of fig tree is pollinated by a different species of wasp. Thus, the extinction of just one or two wasp species could cause several types of fig trees to cease to bear fruit, thus adversely affecting all the animals who eat that fruit.

In addition to playing crucial roles in their native ecosystems, many species provide benefits to humans. In agriculture, wild plants and animals offer possibilities for new or improved crops, while bees that pollinate crops enable them to bear grain or fruit. Plants and animals are also an important source for pharmaceutical products; worldwide, almost half of all medicinal products are derived from living organisms. Medical research has learned from other species as well. For example, the slow-clotting blood of the Florida manatee has aided researchers involved in the study and treatment of hemophilia. Manufacturing industries also depend on biological resources. For example, jojoba, once considered a desert weed, yields a wax that is used as an industrial lubricant in place of sperm whale oil. Significantly, only a small fraction of the world's species have been studied for the role they play in their ecosystem or their potential benefits to mankind. Many will disappear before we even learn that they exist!

In addition to the aesthetic and practical functions different species perform, many people believe that each and every species has a right to exist for its own sake, independent of its value or worth to humankind. They believe humans overstep their bounds as inhabitants of the planet when their activities make it impossible for other species to live here as well. Finally, preserving a diversity of species is essential for the process of evolution to continue; the fewer species that survive, the fewer new species there will be in the future.

The Role of Zoos in Protecting and Preserving Endangered Species

The most important measures that can be taken to halt extinction are those that protect and restore the natural habitats of species. The interconnections among species in their natural habitats are so complex and so poorly understood that only in that undisturbed habitat can the full diversity of species continue to exist. In addition, by preserving habitat, we preserve even species that are still unknown to us or that we might not have room for in zoos (U.S. zoos can probably preserve self-sustaining populations—150 individuals—of only 100 mammal species). However, when a species or its habitat is so endangered that it is close to extinction, zoos can play a crucial function in preserving the species in the hope that it can someday be returned to a safe natural environment.

The St. Louis Zoo makes a substantial contribution to preserving endangered species through breeding and research. In recent years, the zoo's staff has made important advances in breeding many types of animals, including the rare lesser kudu (a type of antelope), and the vanishing black rhinoceros. Perhaps the zoo's greatest success has been its unsurpassed record for breeding the endangered black lemur, a small primate from Madagascar.

SOCIAL BOUNDARIES: THE IMPORTANCE OF PUBLIC SUPPORT

The St. Louis Zoo has always enjoyed a great deal of community support - a vital ingredient in the success of any zoo. In 1913, the city set aside 77 acres in Forest Park to establish a zoological park. By 1916, the citizens of St. Louis had approved a tax to finance the construction of the Zoo, making St. Louis the first city ever to support their zoo with a mill tax.

Almost eight decades later, the tradition continues. Not only does a tax paid by city and county residents continue to fund the zoo, but support also comes in the form of individual donations, civic groups, and volunteers. The Zoo Parent Program allows visitors to adopt an animal by making contributions that go toward its care and feeding expenses. Over 9000 parents participate, receiving certificates of parenthood and invitations to the annual Parents Picnic. Zoo-mmm is a zoo-sponsored organization for children. Each child receives the opportunity to participate in a variety of programs that encourage interest in the zoo and exotic animals in general. The St. Louis Zoo Friends Association, a civic group that also works to support the zoo, currently contains about 3000 members.

The zoo also owes its success to the efforts of over 300 volunteers, who supplement the work of the approximately 400 employees. The most highly trained volunteers are docents, who work to educate zoo visitors. They lead animals around the zoo and answer questions, guide school groups around the exhibits, and conduct programs outside the zoo. Other volunteers help with clerical positions, gift shops, information booths, and special events.

LOOKING BACK: THE HISTORY OF THE ST. LOUIS ZOO

The first zoo in the United States opened in Philadelphia on July 1, 1874. This zoo and the others that followed existed chiefly to provide entertainment. The overriding philosophy—"more is better"—led zookeepers to gather as large a collection of exotic animals as possible, even if each species was represented by only one or two members. As they grew in popularity and became more numerous, zoos became a considerable drain on wild populations. When an animal died, a replacement was sought from the wild. To obtain a baby chimpanzee, for example, five or six adults were often killed. Once an animal was captured, there was no guarantee that it would survive the trauma of a move to a foreign area. Those that successfully navigated the journey often succumbed to native diseases and pests against which they had no defenses.

Unfortunately, the trauma of the journey was only the beginning of the animals' misery. Once in their new "homes," they were usually kept in small, dark cages, which were typically poorly vented and nearly always heavily barred. Little or no attention was paid to the animals' psychological needs for comfortable and interesting surroundings, stimuli, companionship, and perhaps most critically, the opportunity to pursue normal behaviors (such as the need of wolves to scent-mark posts or trees).

Like most zoos, the St. Louis Zoo began as a way to entertain visitors with exotic animals. However, modern zoos no longer see their primary role as providing entertainment; instead, they view themselves first and foremost as "arks" for the world's vanishing wildlife. No exception, the St. Louis Zoo has changed its focus to conserving endangered species, and in the last two decades it has succeeded in strengthening the three major areas that support this goal: naturalistic exhibits, breeding and research, and education.

Developing Naturalistic Exhibits

Increasingly, zoos are informally educating visitors via the design of the zoo and its exhibits. The best zoos strive to meet the physical and psychological needs of their animals in a manner that both entertains and educates the public. They do this primarily through exhibiting animals in naturalistic settings that approximate the animals' habitats in the wild rather than in the bare cages once common in zoos. Naturalistic settings provide animals with more stimulation than traditional exhibits by allowing them to pursue their normal behavior patterns.

Before the St. Louis Zoo began major renovations in the 1970s, it contained its share of less-than-natural habitats. For instance, bars separated the great apes from the public, and their cages contained only rubber tire swings and a few tree stumps in the way of amusement. However, the zoo's history also boasts one of the earliest attempts at a naturalized habitat. The bear pits constructed in 1921 featured pools, trees, and walls built to look like rock—all features that resembled the bear's natural environment and gave them more to do and more privacy than a standard, cage-like exhibit.

In the tradition of the bear pits, all of the zoo's recent additions and renovations have emphasized natural surroundings that make the animals feel at home, help stimulate them, and give them privacy. These naturalistic exhibits also aid preservation efforts by giving visitors a sense of the vast scope of ecosystems that support the diverse species of our planet and showing them what advantages and problems each species faces in the wild.

One of the first of these additions was Big Cat Country, an exhibit designed to house lions, tigers, jaguars, pumas, and leopards. The cats' yards are landscaped with rock formations, trees and shrubs, grassy hills, waterfalls, and even pools of water that allow them to swim.

In 1977, the zoo completed similar renovations on the Primate House, which contains groups of monkeys and lemurs. Primate curators have been careful to allow the animals to retain natural family groupings, thus encouraging normal social organization and behaviors. Complete with trees, rock outcrops, grassy floors, and swinging vines, each exhibit also resembles the natural environment of the primates. The exhibits are enclosed with safety glass, protecting the animals from human germs while still allowing visitors to see easily into each area.

The Herpetarium, home to the zoo's amphibians and reptiles (including a 59-year old crocodile, the zoo's oldest resident) was renovated in 1978. The new herpetarium features indoor/outdoor exhibits for some of the larger animals and also provides underground views of some of the aquatic displays. Once again, the zoo simulates the animals' natural habitats as closely as possible, separating visitors from the exhibits with safety glass.

The Bird House, renovated in 1979, uses an especially innovative method to separate birds from visitors. Over 8,300 fine, piano-like tension wires run from the floor to the ceiling in front of the enclosures. Spaced closely enough to prevent birds from escaping, the wires are fine enough that the visitor can focus out the barrier when observing the birds. The wire barrier also provides good ventilation and allows visitors to hear more bird sounds than a glass barrier would. Birds are grouped by zoogeographical regions ranging from rain forests to ocean shores, and these habitats are re-created within the birds' enclosures.

In 1981, with the assistance of the Smithsonian Institution, the zoo added the Living Coral Reef to the aquatic house. An example of a Caribbean coral reef, the display simulates a complete ocean floor ecosystem, from tiny microorganisms to large fish. The aquatic house also contains penguin displays and 21 aquariums exhibiting tropical and native fishes.

One of the zoo's most spectacular habitats, the Jungle of the Apes, opened in 1987, replacing the traditionally constructed Great Ape House. Glass panels separate visitors from the animals, which live in environments designed to mimic their natural rain forest homes. The visitors' trail leads them first into the chimpanzee exhibit. The chimpanzees often take a break from climbing among rocks and trees to fish for honey in their artificial termite mound just as, in the wild, chimps will use long sticks or straw to penetrate holes in termite mounds and draw out the tasty inhabitants. In addition to amusing visitors, naturalistic additions like the termite mound helps watchers learn about the interdependence of species and the intricate relationship between multiple species and their environment.

Next, the trail enters the second exhibit, where visitors are brought eye-to-eye with orangutans by an elevated jungle bridge with a thatched roof. Existing solely on fruit, the orangutans live almost exclusively in the treetops. In the third exhibit, gorillas wander through large trees and mounds of rock. The exhibit allows these huge vegetarians to form a family group as they would in the wild, led by a dominant male called the silverback.

With the completion of the Jungle of the Apes, the zoo achieved an important goal: no animal was housed exclusively behind bars at the St. Louis Zoo.

Aiding Conservation Through Breeding and Research

The St. Louis Zoo employs an expert staff of animal curators, veterinarians, and a reproductive physiologist who all contribute to developing successful breeding programs for a variety of species, most of which are endangered in the wild and rare in captivity. In addition to running the specific breeding programs, they participate in a number of interzoo programs that combat inbreeding, engage in nontraditional breeding methods, and conduct research and other projects that aid animals' health and conservation.

Breeding Programs and the Problem of Inbreeding. As a species' population dwindles, fewer and fewer individuals are available as mates for each other. Consequently, they begin to inbreed, or mate with individuals to whom they are closely related and therefore genetically similar. Thus, the population's variety of genetic traits dwindles as well. Even if the population eventually regains its former size, it will be less genetically diverse. This genetic erosion makes the species even more vulnerable to extinction because it can diminish individuals' fertility, resistance to disease, and competitive ability. Accordingly, genetic erosion is of particular concern to zoo curators and others who manage species' populations.

To combat the problems of inbreeding, in 1973 the Minnesota Zoo instituted the International Species Inventory System (ISIS), a computer-based information system for wild animals species in captivity. As of 1987, ISIS contained basic biological information, such as age, sex, parentage, place of birth, and circumstance of death, for the animals in 220 zoos. This information is used to compile various reports and to analyze the status of captive populations. More importantly, it allows zoos to cooperate in the genetic and demographic management of their animals.

Another program that helps zoo breeders to ensure maximum genetic diversity is the Species Survival Plans developed by the American Association of Zoological Parks and Aquariums. Using ISIS data, the association has developed plans for approximately 40 threatened or endangered species. These plans indicate which animals should mate to preserve the maximum genetic diversity of the entire captive population across zoos.

To help avoid inbreeding in their own and other zoos' animals, the St. Louis Zoo participates actively in the International Species Inventory System, the Species Survival Plan, and the Studbook Program. They even act as international studbook keeper for the black lemur, an endangered species for which the zoo holds the country's best breeding record: over 75 births in the past 20 years.

One of the zoo's most important breeding programs is directed at the Speke's gazelle, an endangered species. Because the zoo could acquire so few of the gazelles, inbreeding posed an inevitable problem. However, zoo staff helped to design an intricate program of genetic planning by which they could breed a small herd and still avoid problems associated with inbreeding.

The zoo is also home to five tuatara, the only members of the endangered species from New Zealand to reside in a U.S. zoo. The exotic, lizard-like reptile has never before reproduced in captivity, and thus the zoo is concentrating exclusively on breeding it. Consequently, visitors cannot even view the tuatara, which dwell in tuatarium, built to resemble their native habitat in the basement of the Herpetarium. Research has led zoo staff to suspect that the female tuatara becomes fertile only once every six years, a disappointing but important discovery.

Other significant efforts to breed endangered species have focused on the Malayan tapir, the bataleu eagles from sub-Saharan Africa, and the Humboldt penguins, which are threatened by the destruction of their habitat (on the west coast of South America), oil slicks, and fishing nets.

Nontraditional Breeding Methods. In some cases, the zoo enhances its breeding efforts through nontraditional methods. One such method, artificial insemination, can help the chance of reproduction among hard-to-breed species and also overcomes the logistical problems of breeding, such as how to bring together two black rhinos, which weight thousands of pounds each, in order to mate. Even more important, the technology can be used to fertilize females with sperm obtained from wild animals, thus rejuvenating the genetic variability of the captive population without disrupting the wild population. In 1980 one of the zoo's Speke's gazelles produced a fawn through artificial insemination.

An embryo transfer can be used to help slow-breeding animals multiply more quickly. Hormones cause the female to produce several eggs rather than one. If the multiple eggs are fertilized during mating they can be carefully flushed from the mother's uterus and injected into the uterus of a surrogate mother to complete their gestation period. At the St. Louis Zoo, an interspecies embryo transfer successfully resulted in the 1984 birth of a zebra calf to a quarter horse mare.

In the same year, the zoo successfully used a third technique, cryopreservation, to produce an eland calf from a frozen embryo. In cryopreservation, samples of both unfertilized egg and sperm as well as embryos arrested at an early stage of development can be frozen in liquid nitrogen and stored indefinitely. The advantage in preserving frozen embryos is this: a population can be sampled and embryos obtained and frozen which represent the genetic diversity of that population at a given time. In the future, when inbreeding threatens the population's health, the embryos could be used to rejuvenate the breed.

Other Areas of Research and Conservation. Other areas of research at the St. Louis Zoo help animals remain healthy after they are born. One such effort began when one of the zoo's black rhinos mysteriously died in 1981. She had suffered from hemolytic anemia, a condition in which the animal's red blood cells seem to self-destruct. After inquiring among zoos worldwide, the zoo found that this problem affects a disproportionate number of black rhinos. Consequently, the zoo formed a research team to determine the causes of this condition and how it can be prevented.

Some of the zoo's conservation efforts entertain and educate visitors at the same time. The Jungle of the Apes, immensely popular among visitors, provides a safe breeding ground for disappearing species while simulating the natural environments of gorillas, orangutans, and chimpanzees, which include rain forests in Western Africa and Southeast Asia. Conversion of the apes' natural habitats to farmland and development is the greatest force threatening their survival, and zoo officials hope visitors will leave the exhibit with a new awareness of the urgent need to save the apes' natural home as well as to preserve the species in captivity.

The Cheetah Survival Center, established in 1974, makes breeding this notoriously hard-to-manage cat a priority while still exhibiting the cats and their offspring. Although the staff has had to overcome various problems (such as the fact that the continuous presence of male cheetahs irritates females and can even prevent them from becoming fertile), the center has successfully bred seventeen cubs so far.

Educating and Entertaining

Zoos present the perfect opportunity to blend education with entertainment, and the St. Louis Zoo takes advantage of every chance to make learning fun. This emphasis on entertaining while educating is for the zoo's benefit as well as that of the visitor; the visitor must understand and support the zoo's goals in order to provide it with the financial and moral support it needs to succeed. In addition, educating the public about the importance of protecting endangered species and their native habitats is an essential complement to the zoo's primary purpose of breeding those species. Besides educating visitors through well-marked and realistic exhibits, the zoo runs a variety of programs designed to appeal to all ages.

The Charles H. Yalem Children's Zoo targets the younger visitors, although many adults also enjoy its displays of smaller animals, baby animals, and species native to Missouri. Other exhibits instruct visitors about animals' lifestyles, including a child-size spider web and a tree-height walkway displaying tree-dwelling animals.

The Children's Zoo shares animals with a much bigger educational center—the Living World, a $17 million teaching facility opened in 1989. The Living World uses live animals in conjunction with state-of-the-art technology to educate visitors about animal diversity, ecology, and conservation. The center contains a theater featuring wildlife films and two major exhibit halls: The Hall of Animals and the Ecology Hall.

Visitors to the Hall of Animals are greeted by a robot of Charles Darwin, who introduces his theory of evolution and the "great family of life" represented in the hall. The hall itself consists of a tour of the animal kingdom from one-celled organisms to mammals. Computer displays, videos, and live animals help bring the tour to life.

The Ecology Hall focuses on the relationship of animals, including humans, to the environment. This hall includes a 60-foot-long model of a Missouri Ozarks stream containing native fish, amphibians, and reptiles. The hall also features videos, weather satellite stations, computer question-and-answer programs, and interactive videos. One interactive video simulates the results of altering a particular environment, such as cutting down a rain forest.

For more traditional education, the Living World also contains one computerized and three standard classrooms that are used for courses that emphasize conservation. Different courses target students at the kindergarten through twelfth-grade levels as well as adults, the disabled, and the elderly. The Living World also includes a teacher resource center and an extensive library. The popularity of the Living World is obvious: annually, over 100,000 children attend courses at the center, which drew over 700,000 visitors in its first nine months alone.

LOOKING AHEAD: THE FUTURE OF THE ST. LOUIS ZOO AND OF THE PRESERVATION OF ENDANGERED SPECIES

The St. Louis Zoo has made great strides in becoming a center for educating the public and breeding endangered species. However, zoo officials still have unrealized plans, such as an elephant house and a small mammal house. The zoo would also like to establish a preservation reserve designed to breed hooved animals, such as the Speke's gazelle, and small mammals. Such breeding preserves play an important role because many animals do not breed well on exhibit because of lack of space and privacy. The preserve would occupy approximately ten acres near the zoo.

Although the St. Louis Zoo's plans will expand its ability to breed and exhibit important animal species, the zoo community agrees that the ultimate purpose of the zoo in today's society must be to protect the diversity of the animal kingdom from the effects of human activities—and no one zoo can accomplish this goal.

The St. Louis Zoo has already taken steps to reach out to other zoos in an attempt to establish some international ties. Bruce Read, the zoo's Large Mammal Curator, traveled to Malaysia in 1989 to teach Malaysian zookeepers techniques for animal conservation and husbandry. He covered topics such as exhibit development, management strategies, parasite screening, and worm schedules. He also demonstrated the use of a computer program designed to process breeding information and developed a breeding program for the Malaysian tapir populations in three of the zoos. Read traveled to Vietnam as well, where he helped build a breeding structure for the endangered kouprey, a species of wild cattle.

The St. Louis Zoo has also established an international alliance with the Beijing Zoo in China. The directors of the two zoos have discussed permanently exchanging certain species, such as the panda, the golden monkey, and certain types of North and South American reptiles and primates.

Because endangered species exist in third world countries and industrialized nations alike, zoos and breeding programs all over the world must share knowledge and animals in order to preserve our planet's biodiversity in the face of human assault on the environment. Those efforts, coupled with efforts to preserve natural habitats and ecosystems, can help to ensure that the Earth of the twenty-first century possesses the beauty and variety that we enjoy today.

SUGGESTED READING

Arrandale, Tom. "A New Breed of Zoo." *Sierra*. November/December 1990. 26+.

"Barn Raising in Vietnam." *ZUDUS* (bimonthly publication of the St. Louis Zoo). July/August 1988.

Balke, Jennifer. "Reproductive Research Programs Offer Future to Endangered Species." *Animal Kingdom*. Published by the St. Louis Zoo.

Begley, Sharon. "A Question of Breeding." *National Wildlife*. February/March 1991. 12-16.

Bendiner, Robert. *The Fall of the Wild, the Rise of the Zoo*. E.P. Dutton, 1981.

Birchall, Annabel. "Zoos Must Join Forces to Save Threatened Habitats." *New Scientist*. June 16 1990. 24.

Bourne, Russell, ed. *A Zoo for All Seasons*. Smithsonian Institution, 1979.

"The Cheetah Chase!" *ZUDUS* . May/June 1987.

Cohn, Jeffrey P. "Captive Breeding for Conservation." *BioScience*, May 1988.

———. "Reproductive Biotechnology." *BioScience*. October 1991. 595-598.

Greene, Melissa. "No RMS, Jungle Vu." *The Atlantic Monthly*. December 1987.

The History of the St. Louis Zoo. St. Louis Zoo Information Series pamphlet.

"Inside the Living World." *ZUDUS* . September/October 1988.

"Jungle of the Apes: A Walk on the Wild Side." *ZUDUS*. May/June 1986.

Luoma, John. "Prison or Ark?" *Audubon*. November 1982.

Read, Andrew F., and Paul H. Harvey. "Population Biology: Genetic Management in Zoos." *Nature*. July 31 1986. 408-410.

"Rhino Report." *ZUDUS*. March/April 1986.

Roberts, Leslie. "Beyond Noah's Ark: What Do We Need to Know?" *Science*. December 2 1988. 1247.

"St. Louis Zoo Cooperates with Scientific Community." *ZUDUS*. January/February 1987.

St. Louis Zoological Park 1986 and 1987 Reports.

Threatened and Endangered Species Exhibited at the St. Louis Zoo. St. Louis Zoo Information Series pamphlet.

Tudge, Colin. "A Wild Time at the Zoo." *New Scientist*. January 5 1991. 26-30.

Woodroffe, Gordon. *Wildlife Conservation and the Modern Zoo*. Hindhead, England: G.W. and Saiga Publishing, 1981.

CHAPTER 21

CULTURAL RESOURCES: THE STATUE OF LIBERTY AND ELLIS ISLAND

When we discuss the environment, we almost always have in mind that vast network of plants, animals, water, soil, air, and other components that we might also call nature. However, another interpretation of the term environment would also include the human species, and everything we produce. Although many of our activities and products hamper the functioning of other parts of the environment, others, such as cultural resources, can play a more positive role.

The cultural resources of a society include both the physical objects its people produce, such as artwork, monuments, and buildings, and their intangible resources, such as language, customs, traditions, and folklore. Cultural resources can be intimately connected to the natural environment. For example, the tribal culture of some native dwellers of the rain forest includes an intimate knowledge of that natural system, a knowledge that enables them to take advantage of the medicinal uses of rain forest plants and to live sustainably off the natural resources available in that threatened ecosystem. In other cases, cultural resources can help a people to maintain a unique sense of their identity, a sense that can also encourage them to preserve their unique cultural and natural resources.

One of the most well-known cultural resources in the United States is the Statue of Liberty. This striking monument, originally named "Liberty Enlightening the World," greeted the immigrant ancestors of roughly 100 million Americans as they saw this country for the first time. Thus, to Americans, the statue represents this country's immigrant heritage. To people around the world, the statue symbolizes the promise of freedom and equality which the United States has held out to so many.

Today, the Statue of Liberty and nearby Ellis Island, where immigrants were processed, are preserved as a National Monument. The history of this monument graphically illustrates how a culturally produced artifact can inspire the imagination, loyalty, and pride of people around the world, and how we must carefully preserve such resources against threats like overuse, vandalism, aging, and pollution if future generations are to share that inspiration.

PHYSICAL/BIOLOGICAL BOUNDARIES: THE STATUE'S STRUCTURE AND ITS DETERIORATION

The Statue of Liberty, rising 151 feet and weighing 225 tons, stands atop Bedloe's Island (now Liberty Island) in New York Harbor. Here, at the gate to the New World, Liberty lifts a torch in her right hand, while her left hand grasps a tablet inscribed with the date July 4, 1776—a commemoration of the Declaration of Independence. Broken chains, representing freedom from slavery, lie at Liberty's feet, and the heel of her right foot is lifted as in walking. The seven rays on her crown symbolize the seven seas, the seven continents, and the seven planets known at the time of her construction. The rays range up to nine feet in length and weigh up to 150 pounds each. The monument rests on a granite pedestal 89 feet high, which in turn rests on a 65-foot foundation. The statue towers a total of 305 feet and 1 inch from the foot of her base to the tip of her torch.

Frederic-Auguste Bartholdi, the statue's French creator, began with a small clay model, then enlarged the model in plaster three times. Three hundred copper sheets were carefully hammered around the final mold using the ancient art of repousse, a technique that caused the thickness of the copper skin to vary from 3/32 inch to 4/32 inch (1/8 inch). This sculptured skin, which measures 11,000 square feet and is approximately the thickness of cloth drapery, serves a structural purpose in addition to its aesthetic beauty. The hammered copper forms a rigid envelope, while the many folds of Liberty's robe distribute stress and minimize sagging.

Bartholdi's countryman, Alexandre-Gustave Eiffel, who would later design Paris's Eiffel Tower, devised an interior iron armature to support the statue. This "skeleton" was composed of over 1800 2 inch by 5/8 inch iron straps that conformed to the inner surface of the copper skin. The straps were supple enough to allow the skin to move in response to winds and thermal stresses. Copper saddles (pieces of U-shaped metal) and copper rivets secured the iron straps to the skin. Flat iron bars connected the straps to a stronger interior structure, which was also made of iron. This structure was connected directly to the statue's interior iron pylon. Four legs, which ran through the pedestal and into the ground, anchored the pylon and provided a secure base for the statue.

Eiffel and Bartholdi were aware that copper and iron will react if they contact each other in the presence of an electrolyte such as water. They tried to prevent this galvanic reaction by inserting shields of asbestos cloth soaked in shellac at every junction between the two metals. But the shellac's ability to prevent the movement of water through the cloth was short-lived. The asbestos alone could not prevent this movement, and eventually probably accelerated it because of capillary action, the tendency of liquids to be drawn into narrow tubes and the tiny openings of porous material. Consequently, as saline water entered the statue's interior through various openings, including rivet holes and the joints of the copper sheets, it acted as an electrolyte. The water caused the iron bars of the statue's armature to corrode everywhere that they contacted the copper skin and saddles.

The rust damaged the structure in several ways. As it ate into the bars, it progressively weakened the iron. And, as the rust accumulated, it occupied a greater amount of space under the saddles than had the original iron bars. The slow but powerful expansion of the rust pulled the rivets that secured the saddles right through the copper skin, detaching many of the saddles and causing holes in the skin which admitted more water, further hastening the deterioration. By the 1980s, a substantial number of the bars had lost as much as half their original thickness.

Another problem contributing to the statue's degradation arose from alterations made to the flame of the torch. Originally, Bartholdi had fabricated the torch flame from solid sheets of copper. But prior to the statue's dedication in 1886, two rows of portholes were cut into the lower half of the flame so that it could be illuminated from inside. Because this illumination was dim, a band of glass was installed in 1892 to replace the upper row of portholes, and a pyramid-shaped skylight was inserted at the top of the flame. Then, in 1916, the entire surface of the flame was altered into panes of amber glass in a copper framework. An elaborate glazing system, devised to prevent water from entering the interior through the mosaic-like flame, failed. Hence, the torch became the statue's single worst source of water penetration, and the leakage and resulting corrosion severely damaged both the torch and the upraised right arm.

Caretakers of the statue tried to combat the leakage from the torch as well as from stretching seams and rivets, but to no avail. In 1911, they applied a bituminous paint (coal tar) to the interior of the statue in an attempt to isolate the iron and copper from each other. As the paint peeled and blistered, it trapped moisture inside, further complicating the problem.

SOCIAL BOUNDARIES: A SYMBOL OF THE UNITED STATES AND THE IMMIGRANT TRADITION

The Statue of Liberty, erected in 1886, has always served as a potent symbol of values important to the United States and to the people of France who gave us the statue. It represents the ideals of liberty and freedom from oppression; justice and equality before the law; the struggle for American independence; the ideal of American democracy; the American heritage of freedom, hospitality, and welcome; opportunity; hope for the future; and the promise America represents to the world.

Although the statue was carefully constructed as a monument to freedom and equality of opportunity, it was its position in New York Harbor which allowed it to come to truly embody these ideals. The harbor also contains Ellis Island, which became America's official immigration station in 1892. A total of 17 million immigrants passed through Ellis Island, and over one million passed through in the peak year of 1907 alone.

For many of those immigrants, the memory of arriving in their new home was inextricably linked with the image of the statue lighting their way into the Harbor. Perhaps the significance of the Statue of Liberty to the world of the late 1800s and early 1900s can best be summed up by the poem, "The New Colossus," which was engraved on its pedestal. (The title and the first two lines of the poem refer to the Colossus of Rhodes, a bronze statue of approximately 120 feet which stood near the harbor of Rhodes on the Aegean Sea in the early 200s B.C. The Colossus was considered one of the Seven Wonders of the Ancient World.)

The New Colossus
by Emma Lazarus, 1883

Not like the brazen giant of Greek fame,
With conquering limbs astride from land to land
Here at our sea-washed, sunset gates shall stand
A mighty woman with a torch, whose flame

Is the imprisoned lightning, and her name
Mother of Exiles. From her beacon-hand
Glows world-wide welcome, her mild eyes command
The air-bridged harbor that twin-cities frame.

"Keep, ancient lands, your storied pomp!" cries she
With silent lips. "Give me your tired, your poor,
Your huddled masses, yearning to breathe free,
The wretched refuse of your teeming shore.
Send these, the homeless, tempest tost, to me;
I lift my lamp beside the golden door!"

Despite the thrill and opportunity of starting a new life, the passage through the golden door was not an easy one. To a newcomer in a strange land, the immigration process was always difficult and often frightening, consisting of many long lines, impersonal interviews, and quick medical inspections. Known as "The Island of Tears," Ellis Island is nonetheless an important symbol, representing the courage and perseverance of the immigrants, the spirit of sacrifice, hard work, new beginnings, and tolerance for ethnic and cultural diversity.

Because the island increasingly came to be used for deportation after immigration restrictions became tighter in the 1920s, its history also shows that our country's doors are not open to everyone. The immigration station was finally closed in 1954, and in 1965 President Lyndon B. Johnson declared Ellis Island part of the Statue of Liberty National Monument. Together, the statue and the island are visual reminders of our immigrant heritage.

LOOKING BACK: THE HISTORY OF THE STATUE OF LIBERTY AND OF ELLIS ISLAND

Although the statue has always been an important monument to the ideals of this country, its history along with the story of Ellis Island demonstrates that we must always be vigilant in protecting our cultural resources from the damage caused by human use and environmental processes.

How the Statue Came to the United States

In 1865, the United States had just survived its most critical trial to date—the Civil War—and was emerging from the blackest period in the young nation's history. Although President Abraham Lincoln had recently fallen victim to an assassin's bullet, the goals he had worked for were achieved: the union was preserved and the abolition of slavery represented another step toward the ideal of freedom for all men.

The events in the United States were widely discussed in France, which was then laboring under the oppressive regime of the Emperor Napoleon III. At a dinner party in 1865, Edouard de Laboulaye, a French historian and legal scholar, proposed that France give the United States some sign of the long friendship between the two countries—a monument dedicated to their shared ideals of freedom. Bartholdi, a guest at the dinner, enthusiastically embraced the idea, envisioning the monument as a robed woman holding aloft a torch. Five years passed before the collapse of Napoleon's empire, but when the French reestablished their own democracy in 1873, Laboulaye and Bartholdi were finally able to launch their project.

In designing the statue, Bartholdi took his inspiration from earlier monuments, including the Colossus of Rhodes and the 76-foot copper statue of St. Charles Borromeo in Arona, Italy. He also found inspiration closer to home; the face of the statue is based on the features of Bartholdi's own mother.

Although designing the statue presented some artistic and technical problems, they were inconsequential compared to the difficulty that arose from paying for its construction. The French had few problems raising the $400,000 needed to construct the monument itself, but the Americans had a far more difficult time raising the money to pay for the statue's pedestal.

Although Congress had accepted the gift and approved the Bedloe Island site chosen by Bartholdi, it would appropriate no money for the base. The $50,000 that the New York State legislature approved was vetoed by Governor Grover Cleveland. Fund-raising efforts were unsuccessful because of a lack of public interest and support. When only half the necessary funds had been raised by 1883, the project was in danger of failing. Joseph Pulitzer, an immigrant and owner of several newspapers, including the *New York World*, took up the cause and stimulated support for the statue. He embarrassed the rich for not contributing to the project and stressed that the statue was a symbol for all Americans. He published the names of all contributors—over 121,000 in all.

Finally, on July 4, 1884, the monument was presented to the American Minister in France. The statue was then dismantled and shipped to the United States, where it was reassembled. The Statue of Liberty Enlightening the World was formally dedicated by President Grover Cleveland on October 28, 1886.

How the Statue Was Assembled and Maintained

Unfortunately, U.S. workers who reconstructed the statue sometimes failed to match French rivet holes or plans for the framework. As a result, the head was assembled approximately two feet out of alignment, resulting in unexpected stresses on the monument and causing one spike of the crown to touch the statue's upraised arm and eventually puncture the copper skin. Over the years, attempts were undertaken to maintain the statue, but its large size and waterbound location made these efforts sporadic. Meanwhile, the marine environment, combined with the well- meaning "improvements" to the design of the torch and the many leaks, caused the statue to slowly deteriorate.

In 1924, the Statue of Liberty was declared a national monument; nine years later, it was placed under the jurisdiction of the National Park Service. Still, management efforts were largely "brushfire tactics"—dealing with the worst of the deterioration in a haphazard manner. For instance, at least seven coats of paints were applied to the statue's interior throughout this century in an attempt to prevent the copper-iron reaction. Then, in 1937-1938, the most severely corroded iron bars were replaced.

How the Statue and Ellis Island Were Restored

Finally, in 1982, the NPS realized the urgency of restoring both the statue and Ellis Island in preparation for their upcoming centennial celebrations, 1986 and 1992, respectively. Then-Secretary of the Interior James Watt appointed a Statue of Liberty-Ellis Island Centennial

Commission to advise the government on restoration activities. Lee Iacocca, head of the Chrysler Corporation and the son of immigrants, was named chairman of the commission.

Because the Park Service budget precluded the allocation of any funds to the restoration efforts, a nonprofit private corporation, the Statue of Liberty-Ellis Island Foundation, was formed to raise funds from private contributions. The foundation assumed the responsibility for all fundraising activities connected with the restoration, and also assumed the power to make contractual decisions regarding the restoration work and to oversee the work in progress. Just as the enthusiasm of U.S. citizens and corporations for the statue inspired them to raise the money for its pedestal in the 1880s, their pride in this cultural symbol inspired them to donate a total of over $230 million to restore both the statue and Ellis Island.

Restoring the Statue. Restoration first required a careful analysis of the condition of both the exterior copper skin and the interior iron armature. The copper's green patina, produced by a complex reaction between copper and atmospheric gases in the presence of water, provides a natural shield for the skin, slowing the process of corrosion. Tests demonstrated that the skin was in surprisingly good shape; only 4 percent of its original thickness had corroded in 100 years.

In the recent past, however, darkened areas have appeared on the patina, particularly on the north side facing Manhattan. Corrosion specialist Robert Baboian, head of the electrochemical and corrosion laboratory at Texas Instruments, Inc., in Attleboro, Massachusetts, collected and analyzed scrapings from small areas of the statue's surface. His results suggest that acid rain may be the culprit, converting a stable form of copper sulfate called brochantite, which forms naturally on the copper surface, into a less stable form of copper sulfate called antlerite. Antlerite dissolves more readily in water and is more susceptible to wind erosion. Hence, the antlerite-loaded patina may be washing away, particularly in areas buffeted by the prevailing winds, exposing an underlying black layer composed chiefly of copper sulfide and copper oxide.

Baboian's findings were supported by Park Service studies that mapped the changes that had occurred to the surface over time. Mapping involved careful computer analyses of photographs of the statue as it appeared throughout its first century. By examining photographs taken in the 1950s, the Park Service found a visible loss in greenish patina since that time, particularly on the statue's north side.

Using an ultrasonic caliper, Baboian and E. Blaine Cliver, chief of the Park Service North Atlantic Historic Preservation Center, also measured the thickness of the skin to see if the copper's corrosion rate was changing. They concluded that no one side of the statue was weathering more than any other, and that the copper in both green and dark areas was of the same approximate thickness. Thus, it appears that acid deposition has not yet affected the copper's corrosion rate. Some experts predict that if environmental conditions remain stable, the statue's skin could last for another 1000 years.

The interior of the statue required far more attention. Architects Richard Hayden and Thierry Despont were in charge of this complex operation. They replaced each of the over 1300 armature bars; this skeleton is so intricate that only four bars in any one area could be removed and restored at one time, and only four areas (a total of 16 bars) far apart from one another could be worked on simultaneously. Stainless steel was used to manufacture the bars. Its properties of thermal expansion and elasticity are similar to the puddled iron (much like wrought iron) in the original bars, but it is stronger and reacts only minimally with copper. As an extra precaution, Teflon was placed as a buffer between the bars and any copper surfaces.

Another important improvement involved stripping off the unattractive, damaging layers of paint on the statue's interior. Although liquid nitrogen successfully removed the outside layers, something more abrasive was needed for the bottom layer of coal tar. Bicarbonate of soda, which was eventually chosen, stripped the tar without scarring the copper, but it did have one unexpected and negative effect. When it leaked outside the statue, the bicarbonate of soda reacted with the patina to streak the statue's surface a bright turquoise blue! Fortunately, a water bath and time have almost completely erased the scars from this mishap. A new, simplified stairway built to the top of the statue reveal the freshly cleaned, cavernous interior and impressive skeleton for the admiration of visitors.

The torch and flame had become so severely damaged that they could not be repaired. The supporting armature was completely disintegrated, a condition that allowed the uplifted torch to actually wave in the wind. French craftsmen replicated the original torch and flame, based on precise computer-generated models (photos and a plaster model based on the existing flame were digitized) and covered the flame with gold leaf. Special lights shining from the balcony around the base of the flame illuminate it beautifully.

The plain doors that had provided access to the statue's interior since 1960 were replaced by a new entrance. Twenty-one feet high and nine feet wide, the new entrance consists of two bronze doors featuring ten bas-relief panels of tools and equipment representing the statue's construction and renovation. Within the statue, the high temperatures, which had been known to reach 49° C (120° F) in the stairway, and humidity that had contributed to the statue's corrosion were banished by a new environmental control system. A history of long lines to tour the statue prompted the architects to replace the elevators with a glass, two-level hydraulic lift.

By the centennial celebration, which began on July 3, 1986, the statue was fully restored, ready to hold up its torch torch of freedom for future generations of Americans and immigrants. At the celebration, President Reagan unveiled the restored statue, and 5000 new U.S. citizens took their citizenship oath on Ellis Island while 20,000 more across the country did the same. The celebration included the opening of a permanent exhibit in the second level of the statue's base. The exhibit emphasized the statue's origin and construction and its evolution into a national and international symbol.

Restoring Ellis Island. The restoration of Ellis Island, which began in 1984, continued on after the Statue's centennial celebration. Although all the buildings were not renovated, the main building reopened in late 1990 as the Ellis Island Immigration Museum. The museum's purpose is to give visitors a sense of the immigrant experience as well as historical facts and figures. To help achieve this purpose, the Park Service is developing oral histories and interpretive programs as well as more traditional features.

Within the main building, the Great Hall, an immense chamber with vaulted ceilings where immigrants were processed, has been preserved as an area for meditation. Other rooms feature exhibits of artifacts actual immigrants carried with them, sent by immigrants and the descendents of immigrants in response to an appeal by the Park Service. Artifacts include tradesmen's tools, musical instruments, toys, diaries, and passports. The Railroad Ticket Office behind the Great Hall has been transformed into an exhibit on immigration history titled "The Peopling of America." The exhibit includes special displays on the forced immigration of approximately 11 million Africans to the Americas and the Caribbean through the Atlantic slave trade. The museum also includes the Library for Immigration Studies.

A unique feature that the Centennial Commission plans to complete in time for Ellis Island's centennial in 1992 is a computerized genealogical center. A computer data base on every immigrant who passed through Ellis Island is being compiled based on material (principally the manifests of ships that transported the immigrants) from the National Archives, the Immigration and Naturalization Service and the Temple Balsch Center for Immigration Research at Temple University in Philadelphia. Once completed, the center will allow visitors to look up information on any ancestors who passed through the island. Information will include their country of origin, the ship they came in on, the date they arrived, their physical characteristics, how much money they arrived with, their level of literacy, their intended destination, and any relatives they had already living in the United States.

LOOKING AHEAD: PRESERVING THE STATUE OF LIBERTY AND ELLIS ISLAND FOR THE FUTURE

Although the massive restoration effort has saved the statue from the effects of destructive environmental and human forces, like all resources, it is still subject to renewed or even new threats as time goes on. Therefore, the Park Service must vigilantly monitor the statue's condition in order to maintain the symbol of which our country is so proud.

Although the dark spots on the statue's patina pose no threat to the copper skin at present, increased acid deposition may eventually cause the dark layer to wear away faster than it forms at the underlying copper surface. This process would probably result in a significant reduction in the thickness of the black layer, color changes, and an increased corrosion rate. Because acid rain is a mounting problem in our atmosphere, the Park Service plans to measure the skin's thickness at selected points from time to time. These "spot-checks" will track how quickly the statue's skin is corroding; if the rate begins to increase, the Park Service will be able to take steps to protect the statue.

Other factors that the Park Service should monitor are other forms of air pollution, the corrosive marine environment, and vibrations from low-flying aircraft that could weaken the statue's structure. Admiring visitors could also pose problems for the statue. Because the monument attracts over 14,000 visitors on some days, overcrowding and the noise, vandalism, and structural strain that accompanies it could harm the statue and also detract from the experience of visiting it. In 1989, the Park Service conducted a visitor study that should help them better manage the monument and satisfy visitors in the future. Doing so will help to preserve the statue so that it might continue to inspire future generations.

Although the creation of the Immigration History Museum on Ellis Island is a fitting tribute to the island's historical significance, most of the 27-acre island is still undeveloped and in a state of disrepair. Currently, one option the Park Service is considering is to work with private developers to turn the rest of the island into a conference center.

SUGGESTED READING

Burroughs, Tom. "Liberty Under Repair." *Technology Review.* July 1984.

Cable, Carole. *The Statue of Liberty: A Selective Bibliography of Literature on Her Construction, History, and Restoration.* Monticello, IL: Vance Bibliographies, 1986.

Hall, Alice J. "Liberty Lifts Her Lamp Once More." *National Geographic.* July 1986.

——. "New Life for Ellis Island." *National Geographic.* September 1990. 89-102.

Hayden, Richard S., and Thierry W. Despont. *Restoring the Statue of Liberty.* New York: McGraw-Hill, 1986.

Horn, Miriam. "The Return to Ellis Island: Restoring the Place of 100 Million American Beginnings." *U.S. News and World Report.* November 21 1988.

Kotker, Norman. *Ellis Island: Echoes from a Nation's Past.* Ed. Susan Jonas. New York: Aperture Foundation, 1989.

Matt, Marcia. "Restoring Lady Liberty—An Update." *Materials Performance.* January 1986.

National Park Service. *Statue of Liberty National Monument: Statement for Management.* January 1990.

Pauli, Hertha E. *Gateway to America: Miss Liberty's First Hundred Years.* New York: D. McKay Co., 1965.

Pennisi, Elizabeth. "Salt, Not Acid Rain, May Mottle Ms. Liberty." *Science News.* August 1991 101.

Peterson, Ivars. "Lessons Learned from a Lady." *Science News.* December 20/27 1986. 392-395

Pitkin, Thomas M. *Keepers of the Gate: A History of Ellis Island.* New York: New York University Press, 1975.

"Restoring the Statue of Liberty." *Architectural Record.* July 1984.

"The Return to Ellis Island." *U.S. News and World Report.* November 21 1988.

Rothman, Hal. *Preserving Different Pasts: The American National Monuments.* Champaign: University of Illinois Press, 1989.

Wingerson, Lois. "America Cleans Up Liberty." *New Scientist.* December 25 1986/January 1 1987.